Managing 12 Volts

How to Upgrade, Operate, and Troubleshoot 12 Volt Electrical Systems

Harold Barre

Summer Breeze Publishing
Redwood City, California

Publisher's Cataloging in Publication
Barre, Harold.
Managing 12 volts : how to upgrade, operate, and troubleshoot 12 volt electrical systems / by Harold Barre.
p. cm.
Includes bibliographical references and index.
LCCN: 95-92383
ISBN 0-9647386-1-9

1. Electric power production. 2. Boats and boating—Electric equipment—Handbooks, manuals, etc. 3. Recreational vehicles—Electric equipment—Handbooks, manuals etc. 4. Storage batteries.
I. Title. II. Title: Managing twelve volts.

TK9901.B37 1997 621.31'2
QBI95-20848

Technical and Editorial Support by Bruce Harris

Designed, Produced & Technically Illustrated by
Teutschel Design Services, Palo Alto, California

The installation, operation and maintenance of an electrical system entails an element of risk. Always consult the designer of your system, the manufacture or supplier of your electrical components, all applicable building and electrical codes, and all relevant regulatory agencies before upgrading or troubleshooting your electrical systems. When in doubt seek professional advice. The author, publisher and distributor to this work assume no liability for personal injury, property damage, or loss from using the information in this book.

Summer Breeze Publishing
1017 El Camino Real, Suite 364
Redwood City, CA 94063-1632

Printed in the United States of America

To

Lucy

for her patience

Table of Contents

Chapter 1 Keys to Electrical Self-sufficiency 1

Ocean Racers 2

Cruising Sailboats 2

The Electrical Systems on RVs and Boats Are Similar 3

Three Keys to Electrical Self-sufficiency 3

Understanding Electrical Terms 5

Chapter 2 Determine Your Electrical Requirements 7

Procedure to Determine Your Electrical Requirement 7

Inventory Your 12 Volt DC Devices 8

Determine Each Device's Electrical Requirement 8

Estimate DC Usage 9

Inverters 11

Determining Your Inverter Requirement 12

Inventory Your AC Appliances 12

Determine the Electrical Rating for Each AC Appliance 12

Estimate AC Usage 12

Inverter Size 13
Inverter DC Requirement 14
Inverter Requirements Are Additional to Other DC Requirements 14
Inverter Limitations 15
Worksheet for Determining Your Electrical Requirement 17

Chapter 3 Determine Your Battery Capacity 21
How Batteries Work 22
A Battery Being Discharged 23
A Battery Being Charged 25
Why Batteries Fail 26
Cold Batteries Have Less Energy 27
Batteries Are Inefficient 29
Batteries Self-discharge 30
Why Batteries "Die" and Come Back to "Life" 30
Only One Volt Is Usable Out of the 12 Volts 31
Rapidly Discharging a Battery Decreases Its Capacity 32
A Battery's Life Is Shortened by Discharging It to Less than 50% of Capacity 33
A Discharged Battery Has Difficulty Supplying High Amperage 34
The Charging Voltage Must Be Greater than 13 Volts 36
How Much Current Will a Deeply Discharged Battery Accept? . . 36
How Can You Increase the Current a Battery Will Accept? 37
How Long Should You Charge Your Batteries? 39
Calculate Your Battery Capacity 40

Chapter 4 Types of Lead Acid Batteries 43
Automotive Starting Batteries 44
Maintenance Free Batteries 44
Small Deep-Cycle Batteries 45
True Deep-Cycle Batteries 45
Large Capacity 12 Volt Batteries 46
Gel Batteries 46
Understanding Battery Capacity Ratings 47
Typical Battery Specifications 48
Adding Battery Capacity 49
Batteries in Series 50

Batteries in Parallel 50
Determine the Batteries for Your Situation 52
Battery Maintenance and Safety Precautions 53
Battery Storage 55
Battery Installation 56

Chapter 5 Battery Charging 57
Charging Devices 58
Alternators. 58
Regulating Voltage 59
Automobile Regulators 60
The Problem with Standard Automobile Voltage Regulators 61
Multi-stage Charging 62
Bulk Charging Stage 63
Absorption Stage 64
Float Stage. 64
Equalization Stage 65
Alternators with Multi-stage Voltage Regulators 66
Battery Chargers (Converters). 67
Portable Taper Chargers 67
Constant Voltage Chargers 68
Multi-stage Battery Chargers 69
Inverters with Multi-stage Battery Chargers. 70
Solar Panels 71
Wind Generators. 72
Determine Charger Output Based on Your Battery Capacity 73

Chapter 6 Monitors, Wiring, and Switches 75
Voltmeters 76
Voltage Readings 76
Ammeters 78
Ammeter Applications 79
Electrical Panels 81
Wiring 82
Connectors. 84
Battery Selector Switches 85
Isolators 88
Solenoids 90

Chapter 7 Designing and Operating Your 12 Volt System . . . 93
Keys to Electrical Self-sufficiency . . . 93
Solar and Wind Power . . . 96
Charging Batteries while Operating Other Loads . . . 96
Recommendations for Upgrading 12 Volt Electrical Systems . . . 99
Small RV, Trailer, or Pickup Truck with Camper . . . 99
An RV with an Onboard Generator . . . 103
Small Sailboat . . . 110
Large Sailboat . . . 115

Chapter 8 Understanding Electrical Circuits . . . 121
A Short Course in 12 Volt Electricity . . . 121
Electricity . . . 122
Electrical Current . . . 122
Voltage . . . 122
Resistance . . . 124
Ohm's Law . . . 125
Multimeters . . . 126
Test Light . . . 127
Practice Using a Multimeter or a Test Light . . . 127
Measure DC Voltage . . . 127
Measure AC Voltage . . . 128
Measure Resistance . . . 130
Measure a Light Bulb's Resistance . . . 130
Measure a Fuse's Resistance . . . 131
Measure a Radio, Stereo, or Electric Fan's Resistance . . . 131
Measure a Switch's Resistance . . . 132
Continuity Check . . . 133
Components of an Electrical Circuit . . . 134
Series Circuits . . . 135
A Circuit Is a Circle . . . 135
Parallel Circuits . . . 136
Negative Ground . . . 136
Distribution Panels . . . 138
Identifying Circuit Components on Your Vehicle . . . 138
Power Source . . . 139

Distribution Panel 140
Electrical Loads 140
Conductive Paths or Wiring 140
Practice with a Test Circuit 141
Resistance Determines the Current in a 12 Volt Circuit. 141
Determine Power at a Load 142
Current Measurement 142
Unwanted Resistance. 144
Measuring a Circuit's Voltage. 146
Finding a Voltage Drop 148
Finding the Voltage of an Open Circuit. 148
Problem with the Power Source 150

Chapter 9 Troubleshooting Electrical Circuits. 155
Schematics 155
Reasons Circuits Fail 156
Find a Break in an Open Circuit. 156
Find a Short Circuit 160
Find a Voltage Drop. 162
Troubleshooting Common Parallel Circuits. 162
All Electrical Loads Fail to Operate 163
Failed Lighting Circuits. 164
Failed Electronic Devices. 164
Troubleshooting Charging Circuits. 165
Check a Charging Device's Output 166
Troubleshooting a Battery 167
Inspect the Battery Terminals, Container, and Electrolyte. . . 167
Testing a Battery Using a Hydrometer 169
Open Circuit Voltage 171
Test for a Battery that Fails to Keep a Charge. 172
Load Tester 172
Undercharged Battery 173
Overcharged Battery 174
Determine the Cause of Battery Failure. 174
Electrical Leakage. 175
Troubleshooting an Alternator 177
Troubleshooting Isolators and Diodes 180

Troubleshooting a Solenoid 183
Troubleshooting Battery Chargers 185
Troubleshooting an RV Converter/Battery Charger. 187
Troubleshooting Solar Panels 189
Troubleshooting Wind Generators 190

Glossary 191

Bibliography 199

List of Manufacturers 203

Index 207

Chapter 1

Keys to Electrical Self-sufficiency

A camper, recreational vehicle (RV), motorhome, or boat provides the comforts of home while you are enjoying nature. However, you have to provide all the services that the municipal utilities provide for you at home: water tanks hold the water for the wash basins and shower; waste tanks hold the waste from the toilet and wash basins; propane tanks provide fuel for the oven and stove. All these systems work well. The water tanks provide water for many days, and the propane tanks provide fuel for months.

The problem is replacing the electric company. The storage batteries, which provide electrical power when you are away from a campground or marina, usually have power for the first night, but after the second or third night the batteries sometimes are unable to support the electrical devices.

So you recharge the batteries using a generator that powers a battery charger, or use the main engine that drives an alternator. Unfortunately, even after charging for many hours, the batteries often fail to power the electrical devices.

Is there a solution to this problem? Yes.

OCEAN RACERS

Man has sailed on the oceans of the world for centuries and some men have raced to distant shores—sailing for weeks before reaching land. To be more competitive, sailors have built strong, light weight boats. To make the journey safer, radar, long range radios, satellite navigation systems, computers, and automatic tracking equipment have been added. To make the voyage more comfortable, refrigeration, microwaves, and TVs with VCRs have been installed. Desalinators convert sea water into freshwater, reducing the need for large, heavy water tanks. All of these systems require electricity and lots of it. Some boats have installed solar panels and wind generators to recharge their batteries, but they are not always reliable or able to provide the massive amount of electrical energy required by the sailboat. To solve this problem, the marine industry has designed sophisticated multi-stage battery charging equipment to recharge the batteries quickly but without damaging the batteries. Some of these charging systems are powered by generators; others are powered by the boat's main engine. These charging systems produce the maximum amount of energy to the batteries in a minimum amount of engine running time. These electrical systems must be reliable not only because of the harsh marine environment, but because their failure could have life or death consequences for the crew.

Can you use these charging systems on your RV or boat? Absolutely.

CRUISING SAILBOATS

Hundreds of cruising sailors have equipped their boats with these sophisticated charging systems. You find these people anchored in beautiful bays from Maine to Florida, in the Caribbean and Mexico, or in the harbors of Greece. People just like you and your spouse crew these boats. They also have CD players, color TVs with VCRs, refrigerators with freezers, and microwaves—everything you would expect to find in a well-equipped RV or motorhome. Marinas at resorts are expensive and in some foreign counties nonexistent, so the cruisers spend most of their time at anchor. They must supply all their utilities just like the RVer has to when dry camping. The availability of clean, safe water is a problem in many areas, so desalinators are increasingly common. Propane is available around the world, but recharging the batteries to run all the electrical devices was a problem until the cruisers started using multi-stage charging systems. Fortunately, marine stores catering to long range sailboats usually have the technical expertise to make the average cruising sailboat safe and comfortable.

How do these systems apply to campers, RVs, or motorhomes? RVs do not use radar, desalinators, or other electrical devices found on boats, so why would battery changing equipment used by a boat work on an RV?

THE ELECTRICAL SYSTEMS ON RVS AND BOATS ARE SIMILAR

Some electrical devices, such as radar and desalinators, are not found on RVs, but the electrical systems of RVs and boats are essentially the same. The equipment in the RV looks different from similar equipment in the boat, but it performs the same function. When an RV or a boat plugs into a 120 volt AC outlet at a campground or marina, a converter/battery charger transforms the AC into 12 volts DC, so that the 12 volt devices on board work. Away from campgrounds and marinas, batteries supply the power to the 12 volt devices. RVs and boats have an alternator driven by the engine to recharge the batteries. Most larger RVs and a few boats have generators powering a battery charger to replenish the batteries. Solar panels are found on RVs and boats.

The two electrical systems are fundamentally the same.

When reading this book, think how the information applies to your specific vehicle, and you will see that the information, concepts, and equipment will apply. Also, the information and concepts are applicable to the individual who has a remote home and wants to be self-sufficient from electrical utilities. Examples of alternative energy systems for remote homes are not given, but all the concepts and equipment discussed also apply to this application.

THREE KEYS TO ELECTRICAL SELF-SUFFICIENCY

Your answers to three key questions will enable you to understand what is required to make your vehicle electrically self-sufficient.

1. **How much electrical energy does your vehicle require each day?**
2. **How large should your battery capacity be to support your daily electrical requirement?**

3. **What type of charging device or devices do you need and how much output should be produced to recharge your batteries quickly and efficiently?**

This book helps you to answer each question, and it details how batteries work, the types of batteries and the types of charging devices you can install on your vehicle. For the individual who does not want all the details, the most important points are highlighted. Worksheets and simple formulas are provided to enable you to answer each question.

The first question is the most important; however, the answer varies widely. A small camper or boat may use only a couple of cabin lights, so its daily electrical requirement is small. The owner of a large motorhome or sailboat may use a dozen lights, watch a color DC TV for hours, play a stereo for most of the day, and use an inverter powering a microwave to heat up dinner, so its daily electrical requirement is large. The first question must be answered correctly or the other two questions will be answered incorrectly, resulting in a poorly performing electrical system.

A worksheet at the end of the Chapter 2 helps you to answer the first question: what is your daily electrical requirement?

The answer to the second question determines the battery capacity necessary to support your daily requirement. Unfortunately, a battery is a mystery to most people, but the battery is the "heart" of the 12 volt system.

The lead acid battery people are most familiar with is the starting battery in their car. The starting battery performs this function so well that many people replace their car before they replace the starting battery. When the 12 volt system fails in an RV or a boat, most people blame the battery for the failure, because this mysterious box is the least understood element of the 12 volt system. In reality, it is the most important. The failure may not be the battery but be caused by an imbalance between the daily electrical requirement and the battery capacity and/or an inefficient charging system. Understanding the battery and its relationship between the daily electrical requirement and an effective charging system is critical.

A simple formula at the end of Chapter 3 helps you to answer the second question: how large should your battery capacity be?

Another little understood part of the 12 volt system is how charging devices recharge batteries. Chapter 5 covers different charging devices ranging from alternators and battery chargers to solar panels and wind generators and discusses whether the charging system designed to keep the starting battery charged does an effective job recharging a deeply discharged battery while an RV is dry camping or a boat is at anchor. The chapter discusses the limitations of the battery chargers found on most RVs and boats. Sophisticated multi-stage voltage regulators and battery chargers are presented as an effective way to recharge deeply discharged batteries. The pros and cons of solar panels and wind generators are discussed because they can also play a role in a successfully managed electrical system.

Another simple formula at the end of Chapter 5 helps you to answer the third question: what type of charging device or devices do you need and what output should be produced to recharge your batteries quickly and efficiently?

In Chapter 7, to help you determine what 12 volt systems are appropriate for different vehicles with different electrical and financial requirements, recommendations are made for upgrading electrical systems. In each example, suggestions are made on the type of batteries to install, and which charging system would be the most useful. Also discussed is how the recommended charging processes work, and how you could use various monitors and controls.

Finally, an electrical book is incomplete without a section on troubleshooting. Most of you rely on skilled electrical technicians to solve your electrical problems, which is fine, but a technician is not always available. Other times the problem may be minor, and you may not want to pay the cost of a technician. Chapter 9 has step by step procedures on how to troubleshoot common circuits found on RVs and boats.

Before you can troubleshoot, however, you must understand how electricity and electrical circuits work, so Chapter 8 presents a short course in 12 volt electricity.

UNDERSTANDING ELECTRICAL TERMS

This book presents new concepts and terms, so the following are definitions for a few of the terms used throughout the book. See also Chapter 8, "Understanding Electrical Circuits," and the Glossary.

The 12 volt electrical system found on an RV or a boat is similar to the fuel system on a vehicle and the fuel pump at the gas station.

A fuel pump on the vehicle or at the gas station pressurizes the fuel so that it flows. In this pressurized system, the fuel flow is controlled by the fuel pressure and the size of the hose the fuel flows through. When the fuel pressure is increased, the flow increases. The fuel pressure may be given in pounds per square inch or psi. The flow to your engine could be measured in gallons per hour.

The **voltage** of an electrical circuit is the pressure of the circuit, and the flow or current is measured in **amperes or amps.** Amperes are the amount of charge (electrons) that flows in the circuit per second. The fuel flow is increased if you increase its pressure, resulting in more fuel flowing quickly into the tank. To increase the flow of electricity in an electrical system, you increase the voltage (pressure), thus increasing the amperes or **current.**

The storage tank for your fuel is called a fuel tank, and its capacity is measured in gallons. The larger the fuel tank the more fuel it can hold. A battery is a storage tank for electrical energy, and one way to measure its capacity is in **ampere-hours or Ah.** The larger the battery's ampere-hour rating the more energy it can deliver. As the term implies, time is a factor in determining the ability of the battery to deliver electrical energy. Ampere-hours is the product of the amount of amperes a battery will deliver and the time over which it delivers the amps. A **rate of discharge** of 10 amps for one hour reduces the battery capacity by 10 ampere-hours. A **rate of charge** of 20 amps for 0.5 hours charges the battery by 10 ampere-hours. Several methods are used to determine battery capacity, and they are explained in Chapter 4.

The battery's **state of charge** indicates how much of the battery capacity is available. If the entire battery capacity is available, it is at 100 percent state of charge. If half of the battery capacity is available, it is at 50 percent of charge. Knowing the battery's state of charge is important.

The electrical device that does the work in an electrical circuit, such as a radio, light, or pump, is the electrical **load.** In an electrical circuit if the switch is "on" and the load (radio, pump, etc.) is working, the electrical circuit is a **closed circuit.** If the switch is "off" and the load is not working, the electrical circuit is an **open circuit.**

The battery that starts the engine of an RV or a boat is called a **starting battery.** On some RVs, this battery is called a chassis battery. The battery that powers the electrical loads of the cabin portion of the RV or boat is called the **house battery**. On some RVs, this battery is called a coach battery.

Chapter 2

Determine Your Electrical Requirements

If you do not know your needs, you will never fulfill them.

You need to determine your electrical requirements before selecting your batteries or your charging system. Your vehicle's designer first determined the type of engine and how much fuel it consumes before determining the size of the fuel tank. An automotive designer would not arbitrarily install a 100 gallon fuel tank on a Geo Metro or a 10 gallon tank on a large truck, so why would you install a small battery to supply the electrical needs of a large RV or boat? The daily electrical requirement must be known before you can determine your battery capacity.

PROCEDURE TO DETERMINE YOUR ELECTRICAL REQUIREMENT

Determine your electrical requirement by answering three questions:

1. **What electrical devices are you going to use?**
2. **How much electrical power do they require?**
3. **How long is each device going to be used?**

Inventory Your 12 Volt DC Devices

Go through your vehicle, and list all the DC electrical devices you are going to use. Usually there is only one of each item: one water pump, one TV, or radio. Vehicles, however, have numerous lighting fixtures consisting of one or more light bulbs. If a light fixture has two bulbs and both are normally lit, list the light as two lights because two light bulbs draw twice as much as one bulb.

Determine Each Device's Electrical Requirement

Determine the amount of current each device uses either from the lists below, from labels on the device, or by using an ammeter. (An ammeter measures the current through an electrical circuit and is discussed in Chapter 6.) The device's electrical rating can vary from an insignificant amount (a digital clock might use 0.01 amps) to a substantial amount (an RV's DC refrigerator can use 16 amps or more).

The device's electrical rating is measured in either watts or amps (75 watt light bulb, or a 2 amp electric motor). When adding up all the electrical devices or loads in an electrical system, you can use watts or amps, just as long as you stick with one or the other. Amps is easier to work with because this unit of measurement is used to determine your battery capacity and charging requirements. Most appliances, however, are rated in watts. Fortunately, converting to amps is easy using the following formula:

Amps = watts ÷ volts.

For example: if your AC/DC TV is rated at 54 watts AC and 57 watts DC, it draws 0.45 amps AC (54 watts divided by 120 volts AC = 0.45 amps) and 4.75 amps DC (57 watts divided by 12 volts DC = 4.75 amps). This is an interesting result, especially if you have an inverter. Do you run the inverter to watch TV or plug your TV into the 12 volt system?

After the discussion on inverters, we will look at this problem again.

Table 2-1 shows typical DC electrical devices found on RVs, with their electrical ratings.

Table 2-1 Electrical Ratings of DC Electrical Devices on RVs

12 volt Devices	Amps
Incandescent Lights	1.5-3.0
Fluorescent Lights	0.8-1.7
Radio/Stereo/CD player, each	2-5
Water Pump	4-5
Propane Heater Fan	5-7
Cabin Fan	0.3-1
DC TV	2-7
DC Refrigerator for an RV	16-30
Inverter	?????

Table 2-2 shows typical DC electrical devices found on boats, with their electrical ratings.

Table 2-2 Electrical Ratings of DC Electrical Devices on Boats

12 volt Devices	Amps
Autopilots	1 to 30
Anchor windlass	80-350
Refrigerator	5-8
Depth sounder	0.06-0.5
Masthead light	1.0-1.5
SSB Radio Receive	2
Transmit	20-35
Strobe light	0.8-1.5
Navigation Lights	3-5
Radar	4-7
Desalinator	4-7
VHF Radio Receive	0.6-1.5
Transmit	5-8
Inverter	??????

Estimate DC Usage

Estimate how long you will use each device per day. Express the time in hours or fraction of an hour. Some electrical devices are fairly easy to estimate; others are not. You can make a fair guess on the amount of TV you watch or how long the stereo is on. The question is how do you estimate all those lights? Do you estimate that each light will be on for one half hour? Or do you list each light separately and estimate how long the toilet light will be on compared to the reading light above your favorite chair? The most accurate estimation is to list each light, because some are on longer than others.

Some devices are on only intermittently. A water pump comes on only when a faucet is opened, and an RV's propane heater cycles on only when the temperature drops. A water pump is usually rated at around 3 gallons per minute, so in an hour it would pump 180 gallons. If you do not use this amount of water per day, the pump is on for only a faction of an hour. The fan on an RV's propane heater requires electricity. If the RV is small and the temperature is not too cold, the fan only comes on for a portion of each hour. If the fan requires 7 amps, the fan uses 3 amps in an hour if it is on for 25 minutes each hour (7 amps x 25 minutes or 0.41 hours = about 3 amp-hours).

Once you have completed an inventory of the DC devices, multiply each device's amperage times its estimated usage to get the total requirement. (Don't worry—you just fill in a table at the end of the chapter.)

Certain items on your vehicle are only used on special occasions. For example, an RV may use a propane heater only during cold periods, and a sailboat uses navigation lights only when under way at night. These devices can require large amounts of electricity, but are not always used. To handle these special occasions, you develop a minimum daily requirement and a maximum daily requirement.

Table 2-3 is an example of the daily requirement for an RV with and without using a heater.

Table 2-3 An Example of an RV's Daily Requirement

Device	Number		Rating in Amps		Daily Hours Used		Daily Total in Amp-Hr
Lights	8	x	1.5	x	0.5	=	6
Stereo	1		2		4		8
Water Pump	1		4		0.25		1
DC TV	1		5		4		20
	Minimum Daily DC Total						35 amp-hr
Heater Fan	1		7		3		21
	Maximum Daily DC Total						56 amp-hr

INVERTERS

An inverter uses the energy from a battery to supply power to 120 volt AC appliances, and requires a different set of calculations. If your vehicle does not have an inverter and you have no plans to install one, skip this section and go directly to the worksheet at the end of this chapter.

Inverters do not provide power but instead transform a battery's 12 volt DC to 120 volt AC. The inverter takes the direct current of the battery and switches or chops it into pulses of alternating electrical current. The inverter's transformer takes these pulses and steps up the voltage, so the output of the inverter is 120 volts alternating current (AC). Some inverters produce a crude form of the AC wave, so insure your inverter produces a sine or modified sine wave or some of your applicances will not function properly.

An inverter allows you to operate AC appliances without being connected to a 120 volt power grid or having to start the generator—to run a blender, for example: see fig 2-1. An inverter is great for handling these AC appliances for short periods, as long as the inverter has been properly sized to both the AC appliances and to the capacity of the batteries. Even a high quality inverter will produce unsatisfactory results if you try to operate microwaves, VCRs, toasters, and other AC appliances using an undersized inverter. Also, if the battery capacity is insufficient to support the AC requirement, an inverter will not provide power for long. You must not only understand your AC requirements, but also ensure your battery capacity is the proper size to power the inverter.

Figure 2-1 Inverter Powering a Blender

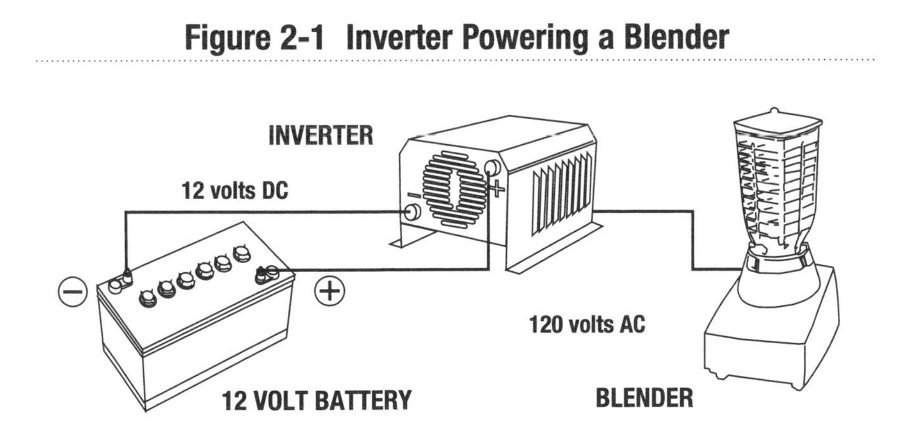

Determining Your Inverter Requirement

You calculate your inverter requirement the same way you determined your 12 volt DC requirement.

1. **What AC electrical devices are you going to use?**
2. **How much electrical power do they require?**
3. **How long are you going to use each device?**

Inventory Your AC Appliances

Make a list of all the AC appliances you expect to use. Blenders, microwaves, hair dryers, computers, and electric drills are all common items that need 120 volt AC, and they can be run using an inverter. A worksheet is available at the end of the chapter to help you to determine your AC requirements.

Determine the Electrical Rating for each AC Appliance

Most AC appliances have labels that indicate how much power they use. Most of them are rated in watts, so for the AC calculation you work with watts. Table 2-4 lists typical loads for AC appliances.

Some AC appliances are not like DC devices, and the differences must be understood. Certain types of AC motors (induction motors like blenders, drills, food mixers) pull three to six times their rated power on initial start-up. This extra power or surge current is required to start the AC motor turning but only occurs for a moment. You may notice this surge of extra current when the lights dim when a large compressor starts on a refrigerator, for example.

Also, electric blenders, frying pans, and incandescent lights require several times their rated load when first turned on. A 75 watt bulb can draw 500 watts when it is first switched on, but the power requirement quickly decreases as the bulb heats. Most inverters are designed to handle this surge current, up to three times the inverter's ratings.

Estimate AC Usage

Estimate the time you expect to use the AC appliances and express the time in hours or fractions of an hour.

Multiply the appliance's wattage by the time you are planning to operate each appliance. The resulting number is the total amount of watt hours each appliance will use in 24 hours. Then total the watt hours for each appliance to determine your total daily AC requirement. Table 2-5 is an example of this procedure.

Table 2-4 Typical Loads for AC Appliances

AC Electrical Appliance	Running Watts
Blender	300-375
Can opener	75-100
Coffee Maker	600-1000
Drill 3/8"	350-500
Frying Pan	1,000-1,200
Food mixer	150-235
Hair dryer	850-1,200
Iron	900-1,200
Microwave	500-1,500
Reading lamp	75-100
Television	50-750
Toaster	800-1600
Toaster oven	1400-1600
Vacuum cleaner	240-1000
VCR	30-50

Table 2-5 An Example of the Daily AC Requirement

Appliance	Watts		Daily Hours used		Daily Total Watt-Hours Used
Microwave	700	x	0.33(20 mins)	=	231
Toaster	1,200		0.17(10 min)		204
Blender	300		0.17		51
Can opener	100		0.1(6 min)		10
Hair dryer	1,200		0.17		204
Coffee percolator	600		0.33(20 min)		198
			Total Daily Demand		898 Watt-hours

Inverter Size

The size of your inverter should be based on the maximum amount of power needed at one time.

In table 2-5 for example, the toaster and the hair dryer each require 1,200 watts, so an inverter larger than 1,200 watts is needed. A 1500 watt inverter would power these appliances. If you wanted to operate both the toaster and coffee percolator (600 watts) at the same time, you would need a 2000 watt inverter.

Inverter DC Requirement

The conversion formula, amps = watts divided by volts, is used to determine the battery capacity necessary to power an inverter. This calculation determines that 75 amp-hours is needed for the example in table 2-5 (898 divided by 12 volts = 75 Ah). Unfortunately, energy is lost as an inverter transforms electricity from 12 volt DC to 120 volt AC. An inverter is only 70 to 90 percent efficient over a range of inverter output. Therefore, it is best to figure the inverter inefficiency at around 20 percent. You calculate this inefficiency, in this example, by multiplying 75 amp-hours by 1.2 and find that you need 90 amp-hours from your batteries.

Inverter Requirements Are Additional to Other DC Requirements

The inverter DC requirement is in addition to the other DC requirements you have on board. Table 2-6 combines the inverter requirement with the requirements of the other 12 volt devices from Table 2-3.

Table 2-6 Daily Requirement Including Inverter Requirement

Devices	Number		Amps		Daily Hours Used		Daily Total Amp-Hours
Lights	8	x	1.5 each	x	0.5 each	=	6
Stereo	1		2		4		8
Water Pump	1		4		0.25		1
DC TV	1		5		4		20
Inverter	1						90
					Min. Daily Total		125 amp-Hours
Heater	1		7		3		21
					Max. Daily Total		146 amp-Hours

In this example, the minimum daily electrical requirement is 125 amp-hours, and the maximum daily electrical requirement is 146 amp-hours.

This brings up the question of the effect an electrical requirement of 146 amp-hours has on your battery capacity. If you have a small 100 amp-hour battery, it will not support the daily requirement of 146 amp-hours. How large should your battery capacity be to support the 146 amp-hour daily requirement? Also how long will it take your existing charging system to recharge your batteries by 146 amp-hours? These questions are discussed in Chapters 3 and 5.

Inverter Limitations

An inverter reduces the battery capacity by a considerable amount if the inverter is allowed to operate for a long period. A small 500 watt inverter reduces the battery capacity by 49 amp-hours if left running for one hour (500 watts divided by 12 volts = 41 Ah x 1.2 efficiency loss = 49 Ah). A 1000 watt inverter drains the batteries of 100 amp-hours if operated for one hour. Even with a battery capacity of 400 Ah, a 1000 watt inverter could drain the batteries very quickly.

Continuously running an inverter is unwise, not because of inverter limitations, but because of limited battery capacity. An inverter should be used to handle small appliances that you need for only a few minutes.

An example is a microwave oven. The small microwaves (500 to 700 watts) are very convenient for heating food, and they do not use much energy—only 12 Ah when heating a bowl of soup for 10 minutes (700 watts x 0.17 (10 min.) divided by 12 volts = 10 Ah x 1.2 inverter inefficiency = 12 Ah). If you cooked something for 30 minutes, you now are consuming 35 amp-hours (700 watts x 0.5(30 min.) divided by 12 volts = 29 Ah x 1.2 inverter inefficiency = 35 Ah).

You may be tempted to install a full size microwave (1200 to 1500 watts) to heat your bowl of soup, but at full power this microwave uses 25 Ah in 10 minutes. The 1200 watt microwave, however, heats the soup in about one half the time. It also requires 75 Ah to cook a dish for 30 minutes. The inverter and microwave work well together to cook a meal in 30 minutes, however, you need to consider what effect removing 75 amps-hours has on the battery capacity. Also, how long is it going to take to recharge the batteries?

The earlier example of the AC/DC TV that draws only 0.45 amps AC but draws 4.75 amps DC is a good example of the inverter inefficiency. If you are going to operate your TV for one hour, would it be more efficient to use the inverter or plug the TV directly into 12 volts DC? At first glance, the inverter seems to be a better choice. But the inverter takes 5.4

Ah of battery capacity to operate the TV versus 4.75 Ah when plugged directly into the 12 volt system (54 watts divided by 12 volts = 4.5 Ah x 1.2 inverter inefficiency = 5.4 Ah). Remember, the inverter just transforms DC into AC. Even though the output of the inverter is 120 volts AC, the inverter is being powered by a 12 volt battery.

Conclusion: always plug directly into the DC when you can.

WORKSHEET FOR DETERMINING YOUR ELECTRICAL REQUIREMENT

Step #1. List all the DC electrical appliances on board your vehicle, determine their electrical ratings, and determine how many hours you plan on using each item. If you are only going to use an item for a fraction of an hour, express the time in decimals (6 mins = 0.1; 10 mins = 0.17; 15 min. = 0.25; 20 min. = 0.33; 30 min. = 0.5; 45 min. = 0.75). After the table is filled out, multiply the number of appliances times the rating in amps. Then multiply this number times the hours each device is used to obtain the total amp-hours used by each device. Then add the total amp-hour column to obtain the total minimum daily requirement of all the DC appliances. (An example is table 2-3.)

Table 2-7 Worksheet to Determine Minimum Daily Requirement

Appliance	Number of Appliances		Rating of Appliance in Amps		Hours Used per day		Total Amp-Hours per day
DC TV	_____	x	_____	x		=	_____
Radio/Stereo	_____		_____		_____		_____
Water Pump	_____		_____		_____		_____
Ventilation Fan	_____		_____		_____		_____
Lights	_____		_____		_____		_____
_________	_____		_____		_____		_____
_________	_____		_____		_____		_____
_________	_____		_____		_____		_____
_________	_____		_____		_____		_____
_________	_____		_____		_____		_____
_________	_____		_____		_____		_____
_________	_____		_____		_____		_____
_________	_____		_____		_____		_____
_________	_____		_____		_____		_____
_________	_____		_____		_____		_____
			Minimum Daily DC Requirement				_____

Step #2. List the special case devices or appliances used only on special occasions. Examples are propane heaters on RVs or navigational lights on sailboats.

Table 2-8 Worksheet to Determine Maximum Daily Requirement

Devices	Number		Rating in Amps		Hours Used Per Day		Total Amp-Hours Per day
______	______	x	______	x	______	=	______
______	______		______		______		______
______	______		______		______		______
______	______		______		______		______
	Total						______
	Add Minimum Daily Requirement (from Table 2-7)						______
	Maximum Daily DC Requirement						______

Step #3. List all the AC appliances you plan to operate with an inverter. Determine each device's rating in watts and how long each will be in use. An example is table 2-5.

Table 2-9 Worksheet to Determine Daily AC Requirement

Appliance	Watts at 120v AC		Hours Used		Total Watt-Hours Per Day
______	______	x	______	=	______
______	______		______		______
______	______		______		______
______	______		______		______
______	______		______		______
	Total Daily AC Requirement in Watt-hours				______

Step #4. Calculate how much battery capacity you need to run an inverter given the above daily AC requirement. Complete the following:

Total Daily AC Requirement _____ watt-hours divided by 12 volts = _____ amp-hours

Due to the inverter inefficiency you must multiply the amp-hours by 1.20.

Amp-hours _____ x 1.20 = _____ amp-hours required to operate the inverter, the inverter requirement.

Step #5: Now take the total daily DC requirement, and add it to the amperage required by the inverter to obtain the total minimum and maximum daily requirement.

Minimum Daily DC Requirement (Table 2-7)	_____	**Max. Daily DC Requirement (Table 2-8)**	_____
Inverter Requirement) (Step #4)	_____	**Inverter Requirement (Step #4)**	_____
Minimum Daily Requirement (Sum of above)	_____	**Max. Daily Requirement (Sum of above)**	_____

You now have the answer to the first question discussed in Chapter 1: How much electrical energy does your vehicle require each day?

Chapter 3

Determine Your Battery Capacity

If you understand how a battery works, you will understand a 12 volt system.

Now that you have determined your electrical requirement, how large should your battery capacity be to support your daily electrical demand? If you were designing a car whose mileage is 12 miles per gallon, you would install at least a 20 gallon fuel tank if you wanted the car to go 240 miles before the car had to be refueled. This same thought process also applies to designing a 12 volt electrical system.

Unfortunately, a battery is not like a fuel tank. The following are some of the differences:

- You can use the fuel to 10% of the fuel tank capacity without effecting the fuel tank. If you regularly drained a battery to 10% of its capacity, you will significantly shorten the battery life.

- A full 20 gallon fuel tank has a usable 20 gallons in it. The miles per gallon change depending on how you drive, but you can use all 20 gallons. A 100 amp-hour battery at 100% of capacity provides only 65 amp-hours if you discharge 65 amps in one hour. The faster you discharge a battery the less capacity the battery provides. The discharge rate is a factor in determining how much battery capacity is available.

- When you pull into a gas station, you refill the fuel tank at the maximum pressure the fuel hose allows. If you recharge a battery to more than 75% at the maximum output of a high output charging device, the battery can become very hot and give off hydrogen and oxygen.
- If you use 20 gallons of fuel, you just refill the tank again with 20 gallons. If you use 100 amp-hours from your battery, it must be recharged with at least 110 to 120 amp-hours. A battery is not very efficient.

As you can see, the process to determine your battery capacity is not as simple as taking your daily electrical requirement and multiplying it by the number of days between recharging to arrive at the proper battery capacity. To determine your battery capacity, you need to learn how the battery works and understand the numerous factors affecting it.

The following sections of this chapter explain how the battery functions and what effect temperature, discharge, and charging rates have on a battery's usefulness. A knowledge of how a battery works is extremely beneficial in explaining why a battery may not always react as you think it should. If you can comprehend how the battery functions, you will understand what type of battery, charging device, and monitors are best for your situation. The rest of the book builds on this knowledge.

HOW BATTERIES WORK

A battery does not store electrical energy but converts chemical energy into electrical energy by electrochemical reaction.

Separate battery plates consisting of active materials react chemically to produce direct current whenever a load (a light, or pump) is connected to the battery terminals. The current is produced by a chemical reaction between the active materials and the electrolyte, a sulfuric acid solution in which the plates are immersed. The active materials are lead dioxide on the positive plates and sponge lead on the negative plates. Separators keep the plates apart but allow the passage of the electrolyte.

Electrolytes contain positive and negative ions, or charged particles. The electrolyte of diluted sulfuric acid contains negatively charged sulfate ions and positively charged hydrogen ions. The chemical reaction of these ions with the plates' active material causes the electricity to flow within the cell. This chemical reaction takes place at the surface of the plates that are in contact with the electrolyte.

The battery's state of charge is measured by reading the specific gravity of the electrolyte. The electrolyte of a fully charged cell has a reading of 1.265 to 1.300 depending on the cell temperature, and whether the battery is designed to be used in cold or warm climates. A starting battery sold in a cold climate has more acid in the electrolyte, resulting in a higher specific gravity reading. A specific gravity reading of 1.26 means that the electrolyte weighs 1.26 times as much as water, which has a specific gravity of 1.0.

A set of plates in the electrolyte is called a cell and has a voltage of about 2.1 volts; see figure 3-1. A battery is formed when two or more cells are connected in series. A 6 volt battery has three cells with a voltage of 6.3 volts. A 12 volt battery has 6 cells, resulting in a voltage of 12.6 volts.

Figure 3-1 Fully Charged Cell

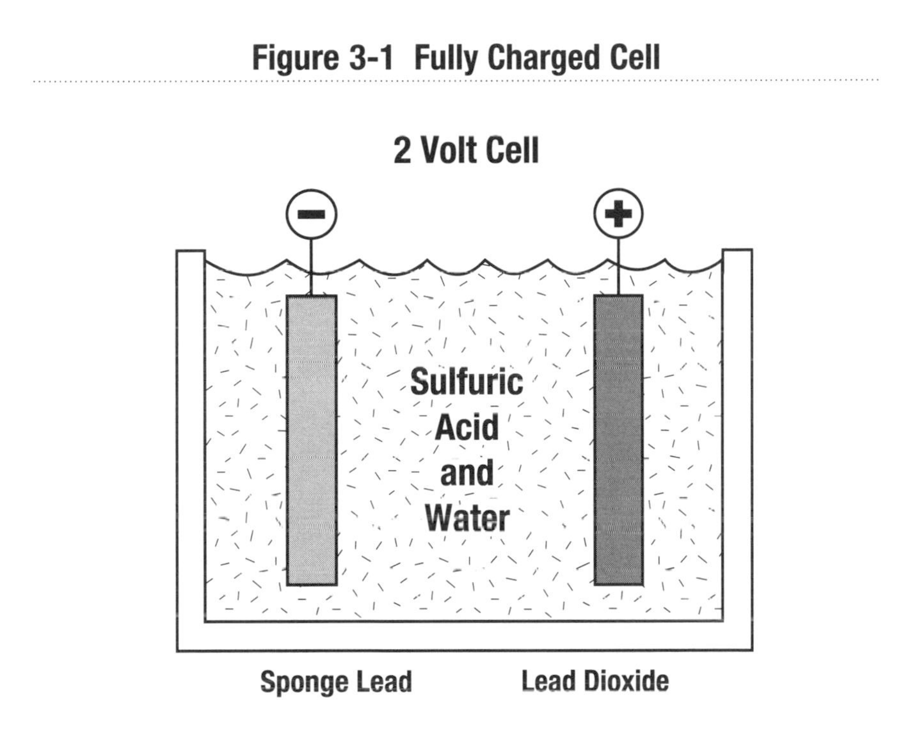

A Battery Being Discharged

When a load (a light or pump) is turned on, figure 3-2, the negatively charged sulfate ions in the electrolyte move to the negative plates where the ions part with their negative charge. This produces an excess of negative charges, called electrons, on the negative plates. This excess of electrons is relieved by electrons flowing through the external electrical circuit to the positive plates.

After the sulfate ions give up their negative charge, the sulfate combines with the lead in the plates to form lead sulfate. The amount of sulfuric acid in the electrolyte is reduced.

The positive plates readily accept the electrons from the external circuit because they are short a negative charge. The hydrogen ions in the electrolyte move to the positive plates and combine with the oxygen to form water, further diluting the electrolyte. The lead on the positive plates combines with the sulfuric acid to form lead sulfate.

The movement of the ions is the source of the cell's electrical power.

During the discharging process, the amount of lead sulfate on the plates and the water in the electrolyte increases. The amount of active material available to react with the electrolyte decreases. The specific gravity of the cell is reduced as it is discharged, making this a convenient way to measure the amount of discharge that has taken place.

At full discharge, the cell voltage falls rapidly because most of the active material has been converted to lead sulfate, and the specific gravity approaches 1.120 because the electrolyte is now mainly water. If the discharge continues, the active material is completely converted into lead sulfate, and the cell can no longer produce sufficient current to power the electrical load. At this point, the cell is completely discharged.

Figure 3-2 A Cell Being Discharged

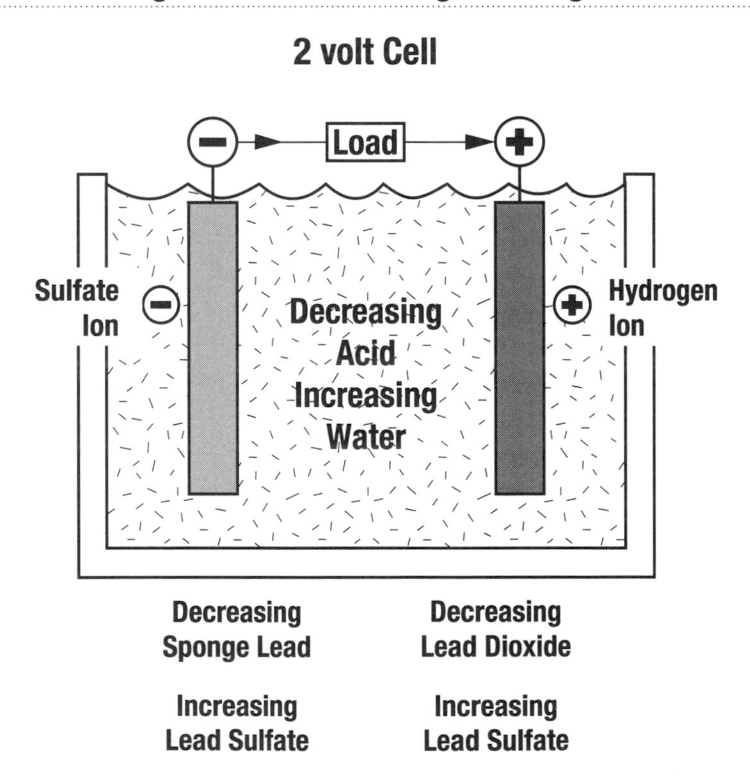

Figure 3-3 A Cell Being Charged

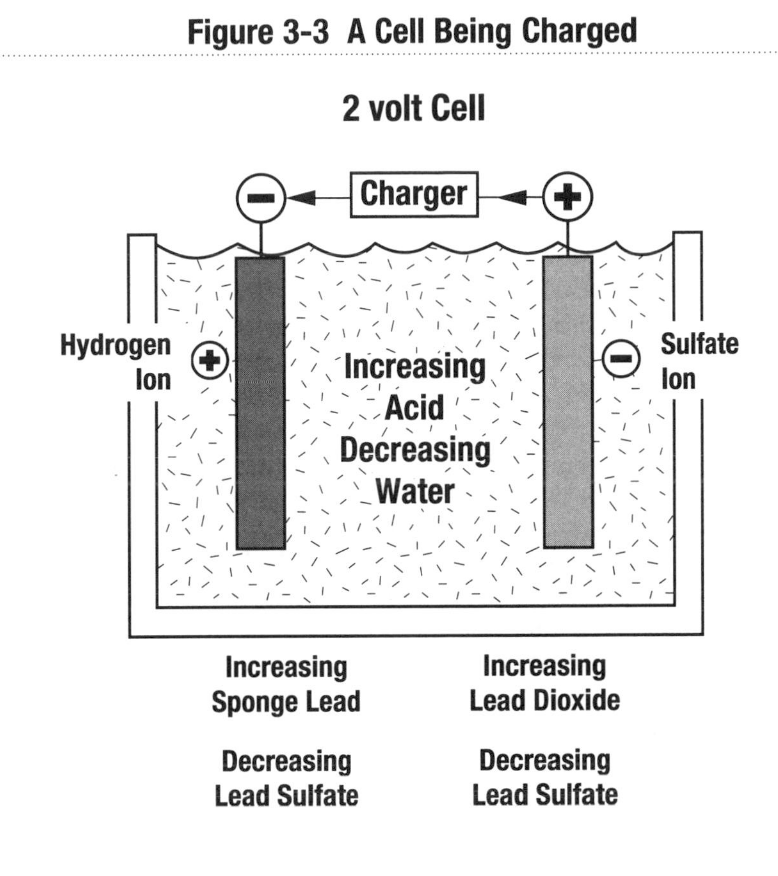

A Battery Being Charged

During charging, figure 3-3, the current is reversed; the charging source has a greater voltage than that of the cell or battery being charged.

The electrical energy of the charging device is converted into chemical energy and stored in the battery.

A charging device provides an excess of negatively charged electrons to the negative plates and creates a shortage at the positive plates. The positively charged hydrogen ions are attracted to the negative plates where the hydrogen combines with the lead sulfate to form lead and sulfuric acid. When most of the lead sulfate is converted to lead, hydrogen bubbles form at the negative plates and rise through the electrolyte.

At the positive plates, the sulfate ions combine with the hydrogen in the water to form sulfuric acid, and the oxygen in the water reacts with the lead sulfate to form lead dioxide. Oxygen appears at the positive plate when this process is near completion. The formation of the gas indicates the cell or battery is nearing complete charge. The voltage at which gassing occurs is called the gassing voltage.

Specific gravity measurements during charging are not a true indication of the battery's state of charge until the battery starts to gas. The gas bubbles mix the stronger acid liberated from the plates with the weaker acid at the top.

At full charge, the specific gravity readings are at their highest, and the plates are again lead dioxide and sponge lead. The cell voltage is 2.1 volts, and the specific gravity is about 1.26 to 1.30, depending on the battery temperature and design.

In Summary:

While discharging, a battery has electrons flowing through an external circuit to the battery's positive plates delivering power to an electrical load. This electrochemical process reduces the acid in the electrolyte and turns the sponge lead and lead dioxide on the plates into lead sulfate, thereby reducing the battery voltage and state of charge.

During charging, the electron flow is reversed. The lead sulfate on the plates is reduced, restoring the plates to sponge lead and lead dioxide. The acid in the electrolyte is increased resulting in the battery's electrochemical power being restored.

Unfortunately, nothing is ever this simple. The batteries will have a long and useful life if they are charged, discharged, and maintained correctly, or they will become a source of frustration.

WHY BATTERIES FAIL

You shorten the lives of your batteries in two ways: by undercharging and conversely by overcharging.

In actual practice, not all the acid is returned to the electrolyte. Some acid remains bonded to the plates in the form of lead sulfate. Lead sulfate is an electrical insulator, preventing the electrochemical reaction from occurring. It is initially soft and relatively easy to reconvert into active material through charging. If a battery is not charged, the lead sulfate crystals grow larger and harden, and they are difficult to reconvert. This reduces the battery capacity. The lead sulfate occupies more space than the active material, and an excessive amount of sulfate places a strain on the plates. If enough sulfate builds up on the plates, the plates could fracture and break off. When a battery is deeply discharged, some of the active material is loosened and falls to the bottom of the battery case; this also reduces the battery capacity. If enough material is shed, it can fill the base of the battery until it reaches the level of the plates and shorts them out—killing the battery. Also, the plate area available for chemical reaction becomes smaller as more lead sulfate remains on the plates. The battery is said to be sulfated or suffering from sulfation. This is the result of not recharging or of undercharging the battery. Sulfation can sometimes be reversed by performing an equalization or conditioning charge on a battery. This procedure is explained in Chapter 5.

Conversely, periods of overcharging or prolonged trickle charging lead to gassing, a process in which excess charging current breaks down the water in the electrolyte into its component parts, hydrogen and oxygen, which then boils off. Water must be added to the cells, replacing the water which boils off. A certain amount of overcharging is necessary to ensure a full charge and the limitation of lead sulfate on the plates, but excessive and prolonged gassing can in time shorten the battery life. Charging is also an oxidation process and as the active material on the positive plates is converted, the positive grid tends to oxidize. With sufficient overcharging, the positive plates oxidize or corrode until they break off, causing the battery to fail completely.

COLD BATTERIES HAVE LESS ENERGY

Temperature is a major factor that affects battery capacity and voltage. Low battery temperature reduces the available capacity and lowers the voltage. Both capacity and voltage are restored with a return to normal temperatures. An increase in battery temperature increases capacity, especially at high rates of discharge.

Capacity and voltage are affected by a change in the viscosity and the resistance of the electrolyte due to a temperature change. At low temperatures, especially below 32°F, the cold electrolyte diffuses through the cell slowly. At high rates of discharge, the battery voltage drops quickly because the electrolyte is slower in circulating to the pores of the active material. Most starting batteries are rated in Cold Cranking Amps (CCA), which indicates their ability to start an engine on a cold winter morning.

At high temperatures, the battery's calculated capacity appears to increase because the electrolyte viscosity and the resistance are reduced. The electrolyte diffuses quickly to the plates, allowing more of the electrolyte to interact with the active material. At high temperatures, the battery is able to support high rates of discharge longer than at lower temperatures. This rapid diffusion causes the battery capacity to increase. This does not mean, however, that the battery can continue to furnish more amp-hours than it received during charge. Unfortunately, if the electrolyte has a continuous temperature over 95°F, the battery life is shortened, and its self-discharge rate increases. As the electrolyte temperature increases to above 110°F, the electrolyte deteriorates, resulting in reduced battery capacity. A battery should never be operated with the electrolyte temperature above 125°F. A battery case feels warm to the touch at 125°F, and if its case feels hot, its internal temperature is at a point where damage can occur.

Low battery temperature temporarily reduces the available capacity, and lowers the voltage. An increase in battery temperature increases capacity, especially at high rates of discharge.

Figure 3-4 Battery Capacity and Temperature

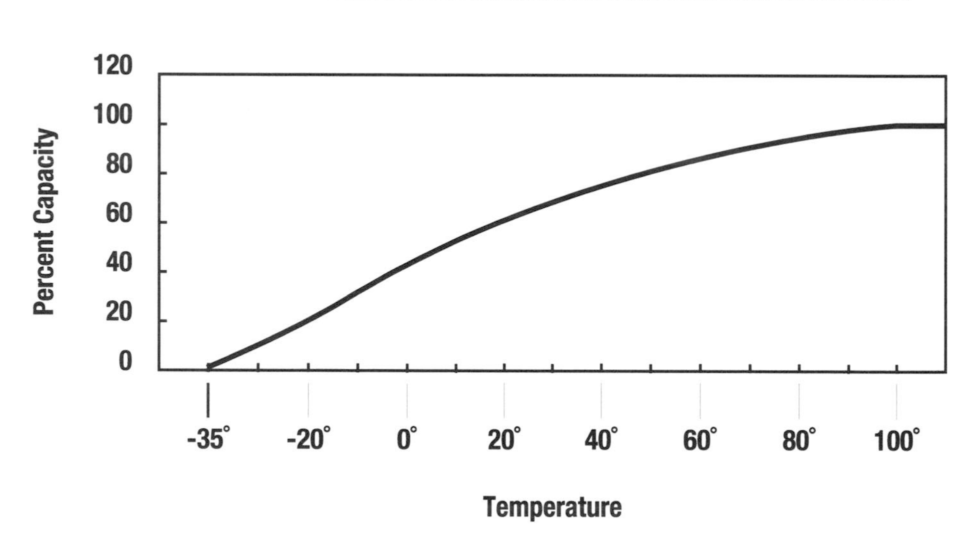

As the graph in figure 3-4 indicates, the battery capacity is reduced as the temperature is decreased. The battery voltage is also reduced as the temperature decreases. The optimum operating temperature for a battery is 80°F.

Little can be done to control the battery temperature, so you should be aware that as the temperature changes the performance of your battery changes. If you are planning to spend long periods in cold weather, your battery capacity needs to be increased. On the other hand, an idle battery in hot weather looses some of its capacity by self-discharging more quickly.

BATTERIES ARE INEFFICIENT

As the battery approaches full charge during the charging process, the battery cannot absorb all the energy from the charging current because almost all of the lead sulfate has been converted back to lead and lead dioxide. The excess charging energy, not being used to reconvert the lead sulfate, breaks down the electrolyte into oxygen at the positive plates and hydrogen at the negative plates. The cell starts to gas freely. This excess charging energy is one of the reasons the electrochemical process is inefficient.

Heat is a byproduct of the electrochemical process. The energy lost to heat during this process is the other reason why batteries are inefficient. Energy is lost to heat in two ways. One is due to the battery's inherent internal electrical resistance. As current moves through the battery, energy is lost as heat. The greater the charge or discharge rate, the greater the energy lost because of the battery internal resistance. The construction of the plates and separators, the purity of the lead and lead dioxide, and the temperature of the electrolyte all determine the battery internal resistance.

The second component of heat loss is due to the nature of the chemical reaction. When the battery is being charged, the chemical reaction within the battery releases heat, but during discharge the chemical reaction absorbs heat. During charging, the battery becomes warm because the battery internal resistance and the chemical reaction release heat. During discharge, the internal resistance still produces heat, but the loss is balanced by heat absorption during the chemical reaction. A battery does not feel as warm on discharge as on charge even though the battery is being discharged at the same rate as when charged. Nevertheless, heat and energy are being lost both during charging and discharging, resulting in inefficiency.

Approximately 10 to 20% additional amp-hours must be returned to the battery to compensate for the inefficiency of the electrochemical process and the generation of heat.

BATTERIES SELF-DISCHARGE

A lead acid battery's state of charge slowly decreases or self-discharges even when no electrical load is connected to the battery.

Temperature, battery age, and type of battery are the primary reasons for self-discharging. As discussed previously, batteries self-discharge more quickly at high temperatures. Self-discharging is reduced at lower battery temperatures. The optimum operational temperature is 80°F, and while it is true that the self-discharge rate would decrease with the lowering of the temperature, the battery capacity also decreases. When storing a battery, place it in a cool area.

The older the battery the more it is susceptible to self-discharging. A battery that is eight to ten years old may self-discharge at two to four times the rate of a newer battery.

True deep cycle batteries have a large percentage of antimony combined with the lead in the plates. The antimony makes the plates stronger, but self-discharging is greater for deep cycle batteries than for other types. A deep cycle battery's capacity can be reduced by 10 to 15 percent per month at 80°F. Batteries that use purer lead in the active material have substantially lower self-discharge rates.

WHY BATTERIES "DIE" AND COME BACK TO "LIFE"

A battery reacts to work like a human—too much output in a short period and it becomes exhausted and needs a short rest. Man, like a battery, can work at a slow pace for hours, but then needs a complete recharge.

The electrochemical process takes place when the electrolyte is in contact with the plate's active material. During discharge, the only useful electrolyte is in contact with the active material. As the sulfuric acid becomes depleted next to the plates, the electrolyte must circulate or diffuse to bring more acid to the plates, or the battery voltage will decrease. The higher the rate of discharge the more rapid this circulation must be to maintain normal cell voltage. As the discharge rate increases, the diffusion does not increase proportionally to the rate of discharge. The electrolyte in the pores of the plates has less specific gravity and is unable to maintain the cell voltage. If the battery is allowed to rest, stronger concentrations of acid from outside the plate area circulate to the pores of the plates so that additional chemical reactions can occur between the sulfuric acid and the lead plates.

At slow rates of discharge, the diffusion of the electrolyte to the plates is continuous. The voltage slowly decreases until complete battery exhaustion occurs. With an extended discharge, all the acid in the electrolyte is consumed. The specific gravity decreases to nearly 1.120, and the voltage decreases to 1.75 volts per cell or 10.5 volts for a 12 volt battery—the final voltage of a 12 volt battery.

A starting battery can deliver 300 amps or more to start an engine, with the voltage dropping to between 8 and 11 volts depending on the battery's state of charge and temperature. This occurs for only a few seconds and does not harm or exhaust the battery. What does deplete a battery to the full extent of its capacity is a discharge at a low amperage for a prolonged period. A battery used to crank a difficult-to-start engine to the point of battery exhaustion comes back to "life" if allowed to rest for a few minutes. However, a battery connected to a small light left on for an extended period will not come back to life; the battery must be charged before it is useful again.

A rapidly discharged "dead" battery comes back to "life" if it sits for a few moments. This permits more acid to reach the plates, thus allowing the chemical reactions to occur.

ONLY ONE VOLT IS USABLE OUT OF THE 12 VOLTS

Approximately one volt separates a fully charged battery from a discharged battery.

A 12 volt battery has a voltage operating range between about 12.6 and 11.7 volts depending on temperature.

However, while the battery is being charged, it may have a voltage reading of 14 volts or higher. During discharge, if a large load is placed on the battery, the battery may have a voltage reading of 11 volts or less.

Figure 3-5 shows that a battery being charged has a voltage of 13 to 14 volts. The same battery at full charge, with no load, has a voltage of about 12.6 volts. When being discharged, the battery voltage decreases to a voltage of about 11.7 where its state of charge is considered to be zero. If a battery continues to be discharged, the cell nears complete exhaustion and the voltage drops rapidly to its final voltage of around 10.5 volts.

Figure 3-5 Battery Voltage

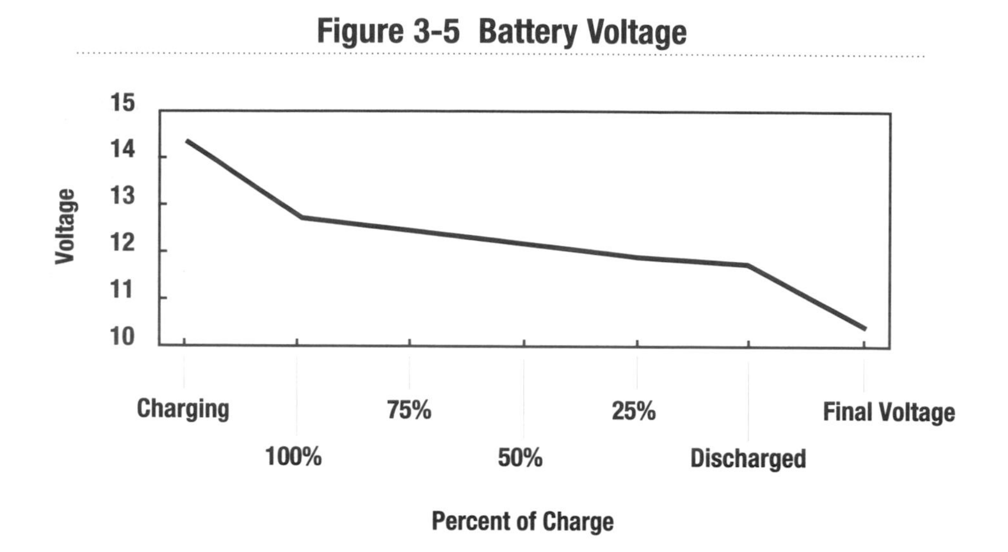

A "surface charge" occurs, after charging, when the voltage remains above 12.6 volts. A heavy load on the battery for 30 seconds can remove this "surface charge" and then the battery voltage can be measured.

You can not determine the battery's state of charge while it is being charged or discharged. The battery voltage must stabilize before the open circuit voltage is an indication of its state of charge. This voltage stabilization occurs after the battery is allowed to rest for at least 15 minutes, preferably longer, after being charged or discharged.

RAPIDLY DISCHARGING A BATTERY DECREASES ITS CAPACITY

The greater the discharge rate, the less battery capacity available.

A 100 amp-hour battery discharged by a continuous 65 amps in one hour is exhausted in one hour, and the battery available capacity is only 65 amp-hours. However, if the battery is discharged at 5 amps, this 100 amp-hour battery provides power for 20 hours. The greater the discharge rate the fewer hours the battery provides power, which results in less battery capacity.

Table 3-1 Capacity of a 100 Ah Battery at 80°F

Discharge Rate	Hours	Capacity
5 amps	20	100 Ah
14 amps	5	70 Ah
65 amps	1	65 Ah

A battery delivers 100 percent of its available capacity if discharged in 20 hours or more, but the battery delivers less than the designed ampere-hour capacity if discharged at a faster rate. A battery that has been discharged at a high rate and appears exhausted continues to provide power if the discharge rate is reduced.

The capacity is drastically reduced at high discharge rates because lead sulfate is rapidly formed at the plate surfaces. The sulfate blocks the pores of the plates, delays and finally prevents further diffusion of the acid to the active material deep in the plates. The battery internal resistance also increases due to the formation of sulfate.

At reduced discharge rates, sulfate is not formed as quickly because the acid's diffusion into the active material proceeds slowly. Thus, a slower discharge rate allows more amp-hours to be available than at higher rates of discharge. The battery voltage decreases slowly instead of dropping rapidly.

A BATTERY'S LIFE IS SHORTENED BY DISCHARGING IT TO LESS THAN 50% OF CAPACITY

As a battery is deeply discharged, the lead sulfate becomes denser and more difficult to reconvert. The battery becomes less efficient and its usable life is seriously affected. The lead sulfate crystals occupy more space than the active material, and if excessive amounts of sulfate are formed, pressure is exerted on the active material and on the grids of the plates. Thin plates are especially susceptible to being destroyed by a build up of sulfate.

An automobile starting battery is not designed for deep discharging, but to provide the maximum amperage output to start an engine. The starting battery's high amperage output is achieved by using a large number of thin plates, which provide ready access to the electrolyte. The plates are thin and porous, so they can provide greater output, particularly at high rates of discharge. A starting battery should not be discharged to less than 60 percent of its capacity, or its life will be shortened.

Deep-cycle batteries, such as 6 volt golf cart batteries, have plates with dense, hard, active material and can be discharged to less than 50 percent of their capacity. But for both starting batteries and deep cycling batteries, if the sulfate is allowed to build up due to deep cycles of discharge and charge, or if any batteries are left for long periods in a state of discharge, their life is prematurely shortened.

As a general rule, a battery should not be routinely discharged to less than 50 percent of its capacity.

A DISCHARGED BATTERY HAS DIFFICULTY SUPPLYING HIGH AMPERAGE

A battery fully charged, in good condition, and operating at 80°F does not have a problem supplying the amount of voltage and current that a load requires. The battery voltage decreases during the normal discharging process or when the temperature decreases. As the sulfate builds on the plates during the normal electrochemical process of discharging or as the temperature drops, the battery internal resistance increases. A battery with low voltage has little problem supplying the electrical requirement of a low amperage load, such as a one amp light bulb. However, when a starting motor requires high amperage and power to start an engine, a battery with low voltage may not supply the needed power.

Following are examples of two batteries supplying voltage to two different loads: one battery has high voltage and the other battery has low voltage.

The first example is a battery whose state of charge is high with an open circuit voltage of 12.5 volts. The two electrical loads are:

A starter motor that requires 225 amps and 2500 watts to start an engine, and

A light bulb that requires one amp at 12 watts.

Table 3-2 shows that a battery with a high state of charge supplies the needed amperage and power for the two examples above.

The power is determined by multiplying the "Current to the load" times the "Voltage at the battery terminals."

Table 3-2 A Battery with an Open Circuit Voltage of 12.5 Volts

Load	Current to the load	Voltage at the battery terminals	Power at the load
Starter Motor	231 amps	11.09 volts	2566 watts
Light bulb	1 amp	12.49 volts	12 watts

You will notice the voltage drops to about 11 volts when the battery supplies high amperage to the starter motor. You can observe this voltage drop when starting an engine because the vehicle lights dim. What causes this voltage drop is the high amount of current through the battery internal resistance. Even through the internal resistance is extremely low, perhaps a few thousandths of an ohm, the high current required by the starting motor flowing through the battery causes a voltage drop, lowering the voltage the battery can supply to the load. This is not a problem for a battery whose state of charge is high.

Table 3-3, shows a battery that is deeply discharged, with an open circuit voltage of 12.0 volts.

Table 3-3 A Battery with an Open Circuit Voltage of 12.0 Volts

Load	Current to the load	Voltage at the battery terminals	Power at the load
Starter Motor	200 amps	10 volts	2000 watts
Light bulb	1 amp	11.9 volts	12 watts

This deeply discharged battery has difficulty supplying enough power to start the engine. The amperage is reduced by only 31 amps from the first example, but the voltage at the battery terminals is reduced to 10 volts, so the power to the starter is reduced to 2000 watts.

The light bulb, because it requires only one amp and 12 watts, is lit and providing light.

The simple answer to this problem is to insure that batteries do not become deeply discharged, and that they are maintained and charged properly.

A deeply discharged battery has a problem supplying current, voltage, and power to a high amperage device, but may supply power to a low amperage device.

THE CHARGING VOLTAGE MUST BE GREATER THAN 13 VOLTS

When charging a battery, the charging device voltage must be higher than that of a fully charged battery.

If the battery and the charging device have the same voltage, no current will flow to recharge the battery. If the battery voltage is higher than that of the charging device, current is from the battery to the charging device and the battery is discharged. If the charging device voltage is higher than that of the battery, current is from the charging device to the battery and the battery is charged.

The greater the voltage difference between the charging device and the battery the greater the current will be. As the voltage of the battery approaches the voltage of the charging device, the current is reduced or tapered, and only a small amount of current (amperage) is accepted by the battery.

Most standard battery chargers charge at 13.8 volts, but some have a voltage of 14 volts or higher. The battery voltage increases as the battery is charging, but the battery's state of charge can not be interpreted from this voltage. The voltage is initially high because the electrolyte around the plates has been restored as more acid is driven back into the electrolyte. However, the electrolyte in the outer reaches of the cell has not been restored; the diffusion process takes time. Directly after charging, the battery voltage is as high as 14 volts but quickly drops as a load is placed on the battery, or slowly decreases on its own. The diffusion process equalizes the concentration of the acid throughout the cell, so that after charging the voltage decreases over time to the battery's true voltage and state of charge.

HOW MUCH CURRENT WILL A DEEPLY DISCHARGED BATTERY ACCEPT?

A deeply discharged battery can absorb a large amount of current, as much as 30 to 40 percent of its capacity, until it starts decomposing water in the electrolyte.

The battery safely accepts this high rate until it is 60 to 80 percent charged. Once the gassing voltage (where the cells start to gas freely) is reached, this high rate of charge must be reduced to the normal absorption rate of the battery. If the charging rate is not reduced, the excess current breaks down the electrolyte into oxygen and hydrogen and overheats the battery.

As shown in figure 3-6, a 100 amp-hour battery accepts 17 amps at 50 percent of charge when being recharged at 13.6 volts. As the battery's state of charge increases to 75 percent, the battery only accepts 7 amps, and at full charge it accepts 1 amp. The current is reduced or tapered as the capacity of the battery is approached.

If a battery is only discharged by a few amp-hours, it accepts only a few amps during charging. If a battery, however, is discharged to 50 percent of capacity, it safely accepts a greater charging rate.

Figure 3-6 Battery Acceptance Rate for a 100 AH Capacity Battery

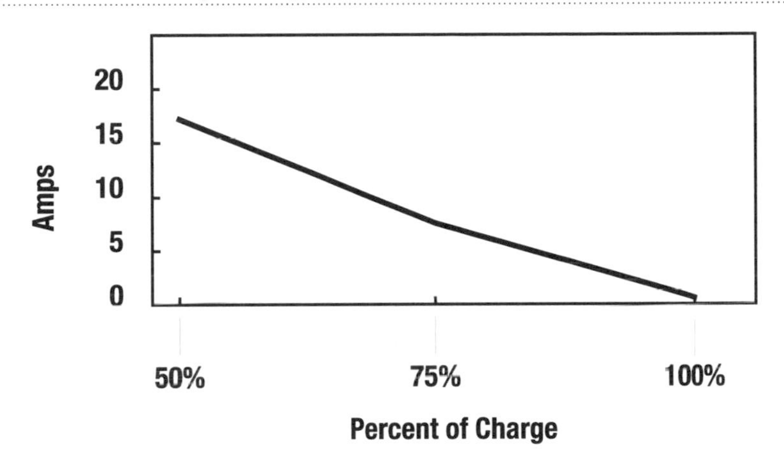

HOW CAN YOU INCREASE THE CURRENT A BATTERY WILL ACCEPT?

The greater the voltage difference between the battery and the charging device, the greater the amount of current will flow into the battery.

Just as a water hose fills a tank faster when the pressure is increased, a battery is filled more quickly if the voltage is increased.

Fig 3-7 shows what happens when the voltage is increased to 14.4 volts for a 100 amp-hour battery.

A voltage increase of only 0.8 volt increases the amperage that the battery accepts at 50 percent of charge to 30 amps, but the amperage that the battery accepts at full charge is only increased to 2 amps. As the battery voltage increases and approaches the charging device voltage, the current is decreased.

By increasing the voltage, you have significantly decreased your charging time.

Figure 3-7 A Battery Acceptance Rate

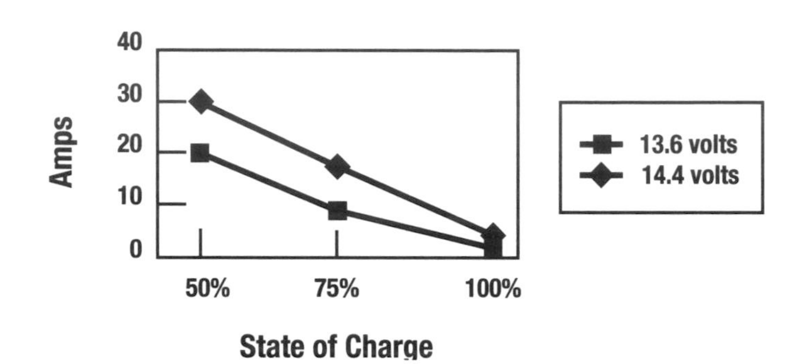

So, this is a great way to recharge your batteries rapidly. You just discharge your batteries to 50 percent of charge and crank up the charging voltage, and you will have your batteries recharged quickly. Right?

Unfortunately, it is not this simple. A couple of other considerations must be understood.

First, if you left the voltage at 14.4 volts, even at 2 amps the charger produces enough energy to decompose the water in the electrolyte.

When the gassing voltage is reached, the voltage and current must be reduced to a low level to prevent the breakdown of the water into oxygen and hydrogen.

Second, the greater the rate of charge, the quicker the battery reaches the gassing voltage, however, the battery's state of charge will be less than if a lower rate of charge had been used.

Table 3-4 shows what happens to a battery that has been discharged to 20 percent of its amp-hour capacity and then is recharged.

A 100 amp-hour battery that is charged with a 10% charging current of 10 amps would reach the gassing voltage in 8 hours and be at 85% of charge. With a 40% charging current or 40 amps, it takes only 1.3 hours to reach 14.4 volts, but the battery would only be at 55% of capacity. The higher the charging rate the quicker the battery reaches the gassing voltage. However, the battery is not as "full" as if a lower charging rate was used.

Table 3-4 Charging Current and State of Charge

Charging Current as % of Ah Capacity	Time to reach 14.4 volts (gassing voltage)	State of charge at 14.4 volts
10%	8 hours	85%
20%	3.5 hours	75%
30%	2 hours	65%
40%	1.3 hours	55%

Again, the problem is the diffusion of the electrolyte. The only portion of the electrolyte that receives the acid is that in contact with the plates. At high rates of charge, the specific gravity in the pores of the plates is enough to support a high voltage, but the acid concentration in the cell's outer reaches is still low; thus the battery's state of charge is low. At lower rates of charge, the diffusion process enables the acid to equalize with more of the electrolyte. It takes longer for the voltage to increase to the gassing voltage, but once there, the battery has a greater state of charge.

The most practical charging rate for a quick recharge is 20 to 25 percent of the battery capacity until the battery reaches the gassing voltage. A rate of charge less than this increases the time for charging; a greater rate of charge results in a lower state of charge.

Once the battery gasses, the current must be reduced to the natural absorption rate of the battery. At this lower rate of charge, additional hours of charging are required, but the decomposition of the electrolyte and heating of the battery are minimized.

When a battery is being charged, the temperature of the electrolyte should not exceed 125°F, nor should there be violent gassing or spewing of electrolyte occurring. If so, reduce the charging rate or stop charging. If the battery case feels warm, the electrolyte is approaching 125°F.

HOW LONG SHOULD YOU CHARGE YOUR BATTERIES?

It is better to keep your batteries fully charged, however, if you charge your batteries while dry camping or at anchor you want to recharge by the most efficient method. You do not want to run your main engine or generator to charge your batteries by only a few amps. You want to minimize the engine running time by maximizing the output of the charging system.

As the graph in figure 3-6 shows, a battery accepts more amperage at 50 percent of capacity than at 75 percent. The battery's acceptance rate tapers to only a few amps as it approaches full charge. If you discharged your battery to only 75 percent of capacity, or 25 amp-hours for a 100 Ah battery, it takes many hours to charge it to 100 percent of capacity. The battery's acceptance rate is only 7 amps at 75 percent of charge and tapers to one amp. However, if you discharged your batteries to 50 percent of capacity, it takes only a few hours to replace the lost 25 amp-hours because the battery accepts 17 amps at 50 percent of capacity, tapering to 7 amps. The 25 amp-hours are replaced relatively quickly. Once the batteries reach about 80 percent of charge, the charging device would be turned off.

You are only utilizing about a third of the battery, from 50 to about 80 percent of its capacity. Enough energy is available to power your electrical devices, to start your vehicle, and to have battery capacity available in an emergency. By charging at 50 percent of the battery capacity, you can quickly and safely replace your daily requirement. You are not running your charging device when the battery only accepts a few amps. Using this method your battery is not fully charged, but you are using the engine or generator in an efficient way.

Periodically you need to bring the batteries to 100 percent of charge. If not, the batteries become sulfated and have a shortened life. The batteries are brought to full charge by the alternator when traveling long distances or by battery charger when plugged into 120 volts AC at a campground or marina. An equalization or conditioning charge may be required to remove the hardened sulfate that has formed. This special charging procedure is explained in Chapter 5.

CALCULATE YOUR BATTERY CAPACITY

Discharging a battery regularly to less than 50 percent of capacity shortens the battery life. Rapidly recharging a battery after it has reached 75 to 80 percent of capacity causes gassing and over heating. Therefore, if you want to minimize the time to recharge your batteries by maximizing the amount of current your batteries will accept, discharge your batteries to 50 percent state of charge and recharge with a current of 25 percent of the battery capacity until the batteries reach the gassing voltage. Therefore, you are only using about 30 percent of the battery capacity.

If you are just using 30 percent of your battery capacity, you need to have 3 times more battery capacity than your daily requirement.

Minimum Daily Requirement______		**Maximum Daily Requirement ______**	
(From the worksheet at the end of Chapter 2)			
x	**3**	**x**	**3**
Minimum Battery Capacity ______		**Maximum Battery Capacity ______**	

If you want to be more conservative, four times the daily requirement could be used to determine your battery capacity. If you decide to use less than 3 times the daily requirements, you need to send more time charging the batteries to bring them to 90 or 100 percent of capacity to ensure you have power for emergency or special situations.

You now have the answer to the second question discussed in Chapter 1: How large should your battery capacity be to support your daily electrical requirement?

If your required battery capacity is much greater than 400 amp-hours, you may want to check the energy requirement of the electrical devices you are operating. For example, 12 volt refrigeration units require massive amounts of energy. Several RV 12 volt refrigerators draw 25 or more amps. If it operates for only 3 hours per day, you need to install at least 225 amp hours of battery capacity (25 amps x 3 hours = 75 Ah x 3= 225 Ah) just to operate the refrigerator. To achieve 225 Ah of battery capacity, you need to install large deep cycle batteries. In warm weather, this refrigerator needs to be operated for more than 3 hours, which requires additional battery capacity. Fortunately, most RV refrigerators also operate using propane, so use the propane and not the 12 volts.

Marine 12 volt refrigerators are much more efficent, but they still require about 7 amps. If it is run for 10 hours it will require about 210 Ah of battery capacity. If you plan on cruising in the tropics and your refrigerator insulation is thin, your refrigerator will not achieve the results you require without adding additional battery capacity. It may be worth looking into an engine driven unit instead of a 12 volt unit.

If your battery capacity requirement seems excessively large, check the electrical devices you are planning to operate and see if you can reduce this need, install more efficient devices, or find other ways to operate them.

Chapter 4

Types of Lead Acid Batteries

If your equipment is wrong for the task, your task will fail.

You have determined your daily electrical requirement and your battery capacity; now, decide what type of battery is best suited for your needs.

To pull a 30 foot trailer, you would not purchase a Geo Metro nor would you purchase an RV for the daily commute through rush hour traffic, so why would you purchase a small capacity starting battery to power the electrical devices on a large RV or boat? Just like vehicles, batteries are designed and manufactured for specific purposes.

The construction of the lead acid battery determines how effective the battery will be for a specific purpose. Where high current is needed to start an engine, but the battery is rarely deeply discharged, battery manufacturers have designed small thin plate batteries that produce tremendous electrical current in a short period—starting batteries.

For a situation where a large amount of electricity is needed over a long period, battery manufacturers have developed large, rugged, thick plate batteries that can withstand repeated deep discharges—deep cycle batteries.

The physical size of the cell determines its capacity, and the more massive the cell is, the greater its capacity. The capacity is increased by increasing the amount of active material and electrolyte contained in the cells. Battery capacity also is increased by making the plates thicker and by adding antimony to make the plates more tolerant of the mechanical stresses of deep discharging and charging. The thicker plates, however, prevent the electrolyte from circulating quickly to the pores of the plates.

AUTOMOTIVE STARTING BATTERIES

Automotive starting batteries are highly developed and specialized to do one thing—start an engine. They are not designed to be deeply discharged and will quickly fail if repeatedly discharged. They should not be used as the house battery on RVs and boats.

The chemical and mechanical construction of a starting battery has been optimized to provide over 300 amps for just a few seconds to start an engine. The battery is then continually charged by the alternator until the motor is shut off. The alternator is powering the automobile electrical system—not the battery. These batteries are constructed with a large number of thin plates that deliver a large amount of amps with a minimum of weight, size, and cost. The large plate area gives the battery the ability to deliver high amperage while maintaining acceptable voltage levels. The thin plates are delicate and disintegrate under the repeated chemical reaction of charge and discharge. Starting batteries begin failing after less than 100 discharging cycles of 50 percent or more of their capacity. They are very poor batteries for deep cycling applications.

MAINTENANCE FREE BATTERIES

Maintenance free batteries are constructed much like starting batteries but have calcium added to the lead plates to harden them and to reduce water loss during charging. The addition of calcium raises the battery internal resistance and prevents rapid charging. The batteries' major advantage is they do not require the addition of water. A sufficient supply of electrolyte has been added to the sealed cells for the expected life of the batteries. With the top of the batteries sealed, it is impossible to check on the electrolyte level or measure the specific gravity. If maintenance free batteries are subject to excessive charging, the electrolyte boils away, causing the cells to dry out and to fail prematurely.

SMALL DEEP-CYCLE BATTERIES

Small compact batteries, group 24 or 27, with thicker plates and some antimony added for additional hardness have been developed to help meet the "deep cycling" needs of recreational vehicles where space and weight are at a premium. The construction of these batteries differs from conventional automobile starting batteries in that the active material is denser, the plates are thicker, and the separators are specially designed to reduce shedding of the active material. The plates are also reinforced, so vibration is less of a problem. They are not designed to power large loads over a long period because they have limited capacity. Small deep cycle batteries are better, however, than normal starting batteries. If they are going to be repeatedly discharged to about 80% of capacity, these batteries usually last for about 200 to 400 discharge cycles. If they are going to be discharged to 20 to 30 percent of capacity, they will last less than 200 cycles.

TRUE DEEP-CYCLE BATTERIES

True deep-cycle batteries are designed to be repeatedly discharged to 20% of their capacity over a period of 5 to 15 years. They are usually of massive size and weight and are rarely assembled into batteries of over 6 volts. The plates are over four times thicker than those in starting batteries, constructed of scored sheet lead, and alloyed with up to 16 percent antimony. The addition of antimony to strengthen the plates, unfortunately, adds to the gassing problems and contributes to a high rate of self-discharge when the battery is standing idle. They may self-discharge as much as 15 to 20% a month. The battery case is larger to allow for additional space under the plates so that shedding of the active material does not cause shorts. Also, more room is available for the electrolyte, providing a reservoir against water loss. The plates are sometimes wrapped in perforated plastic mesh that keeps the lead on the plates longer and extends the life of the plates by as much as 25 to 35 percent. The most common types of deep cycle batteries are built for golf carts and other electric vehicles. They can withstand deep discharging of 300 to 700 cycles.

High capacity, 6 volt batteries are the best type to use for the house battery on RVs and boats.

LARGE CAPACITY 12 VOLT BATTERIES

Large capacity, 12 volt batteries, 4Ds and 8Ds, are designed for starting trucks, tractors, and generators in emergency power situations. Some are designed for deep cycling applications, but many are just starting batteries, so ask before your purchase to insure that the batteries are deep-cycle. The deep-cycle batteries feature the same construction details found in golf cart batteries including multi-rib separators that increase battery life dramatically. The case is large, rugged, heavy and bulky. If maintained properly, these batteries give good service for years.

GEL BATTERIES

Gel, or immobilized electrolyte batteries, have the electrolyte in a gel form and not as a liquid. The tops of the batteries are sealed. The plates are constructed with thin, pure lead or lead-calcium grids separated by special micro-porous fiberglass separators. The plates and separators are compressed, giving the sealed batteries great resistance to vibration. This compression reduces the battery internal resistance allowing them to generate a large amount of current. Because they are sealed units, the batteries must not be allowed to gas. The plates are made with calcium instead of antimony to help prevent gassing. In addition, the negative plates have more active material than the positive plates, so when the positive plates are charged the negative plates are not, and hazardous hydrogen gas is not released.

The natural absorption rate of gel batteries is about twice that of conventional wet batteries. The construction of the batteries with many thin plates results in a large surface area that supports rapid charging and discharging. The batteries are designed for high current, so no current limiting is required for charging. At a constant voltage of 13.8 volts, the natural absorption rate is about 50% of capacity. This rate decreases as the battery approaches full charge, but a charge can be accomplished in 3-4 hours for a deeply discharged battery. Although these batteries can withstand high charging currents, they are sensitive to high charging voltage. Voltages above 14.1 volts at 68°F damage the batteries because they are sealed and gassing must not occur. If they are subjected to high voltage, their warranties may not be honored.

Sealed gel cell batteries are not as sensitive to the effects of repeated deep discharges and will recover 100 percent if left deeply discharged for up to a month. They have a long life due to high purity lead plates, so corrosion at the positive plate is practically eliminated. Since the plates contain calcium, the self-discharge rate is 2 to 4 percent a month. The biggest draw back with gel cell batteries is that they are much more expensive than a comparable wet cell battery.

There is a debate concerning whether quality wet cell batteries are better than gel batteries. One thing in the gel cell's favor is that they are maintenance free, and wet cell batteries need to be monitored for electrolyte level and occasionally may require an equalization charge. A properly maintained high quality, wet cell battery produces far more cycles and gives excellent service for lower initial cost than gel cell batteries. But for someone who wants batteries that do not give off hazardous fumes, require no maintenance and charge more quickly than wet cell batteries, gel cells may be a good buy.

UNDERSTANDING BATTERY CAPACITY RATINGS

Determining the capacity of a battery can be very confusing. Batteries are rated in CCP, CCA, MCA, reserve capacity, peak capacity and amp-hours. Each battery has numbers associated with the rating; some are in amps, and others are in minutes. What does all this mean? Why can't they make it simple like a fuel tank? Again batteries are not as simple as fuel tanks. Temperature and discharge rates influence the battery capacity.

The battery capacity and voltage are reduced as the temperature declines. To start a car on a freezing winter morning, a powerful battery is needed. Manufacturers have standardized a method to measure battery performance for cold winter starts and that is **Cold Cranking Amps (CCA) or Cold Cranking Power (CCP).** CCA or CCP is the maximum discharge current in amps that a new, fully charged 12 volt battery at 0°F can deliver for 30 seconds and maintain a voltage of 7.2 volts. A battery with a rating of 525 CCA x 30 seconds produces 4.3 amp-hours. (525 amps x 0.5 minutes = 262.5 amp-minutes divided by 60 minutes, or 4.3 amp-hours). Since boats usually are not started in zero degree weather, another category called **Marine Cranking Amps (MCA)** has been developed for marine starting batteries. MCA uses the same formula as CCA except the temperature is 32°F instead of 0°F. If you are in the market for a deep-cycle battery, these figures tell you very little, but if you are in the market for a starting battery, **the higher the CCA or MCA rating the easier it will be to start your engine in cold weather.**

Batteries are also rated in **Reserve Capacity,** sometimes called **Peak Capacity.** Reserve capacity is a test in which a discharge load of 25 amps is placed on a battery at 80°F, and the time is measured until the battery voltage reaches an end point at 10.5 volts. A battery with a reserve capacity of 160 minutes has 66.6 amp-hours of capacity. (25 amps x 160 minutes = 4000 amp-minutes or 66.6 amp-hours.) Reserve capacity rating is a good method for determining the battery's suitability as a house battery on an RV or boat. **The higher the reserve capacity the greater the amp-hours a battery holds.**

A 6 volt golf cart battery uses 75 amps as the discharge rate, and not 25 amps. A 6 volt battery with a peak capacity of 107 minutes provides 133 amp-hours (75 amps x 107 minutes = 8025 amp-minutes or 133 amp-hours).

The **amp-hour rate (or 20 hour rate)** is measured at a current that drains the battery capacity in 20 hours to a voltage end point of 10.5 volts. This test is conducted at 80°F. The same battery that provides 66 amp-hours using the reserve capacity rating has a capacity of 100 amp-hours when the 20 hour rating is used. The reason is the 20 hour rating discharges the battery using about 5 amps, and the reserve capacity rating discharges the battery using 25 amps. In Chapter Three, we found that the higher the discharge current the less the amp-hour capacity the battery provides. The same battery powers a 5 amp load for 20 hours, but powers a 25 amp load for only 2.7 hours.

Typical Battery Specifications

Table 4-1 lists the typical capacity, weight and dimensions for various groups of 6 volt and 12 volt deep cycle batteries. Also, gel cell specifications are listed so they can be compared to wet cell batteries. The group number indicates the size of the case.

Table 4-1 Specifications for Deep-cycle Batteries

Group	Amp-hr	CCA	MCA	Reserve Capacity @ 75 amps	Weight lbs.	Length inches	Width inches	Height inches
6 volt Golf Cart Batteries								
	220			107	61	10 3/8	7	11 3/16
	250			149		11 3/4	7	11 1/2
	350			187		11 11/16	7	16 11/16
12 volt Batteries				**@ 25 amps**				
24	85	520	635	135	48	11 1/4	6 3/4	9 1/4
27	105	550	675	160	53	12 3/4	6 3/4	9 1/4
4D	170	787		348	143	20 1/4	8 1/2	9 1/4
8D	216	1,200	1,475	450	164	20 5/8	11	9 1/2
Gel Cell Batteries								
6 volt	180	1,000	1,220	400	70	10 1/4	7 1/8	10 5/8
24	66	400	490	120	54	10 7/8	6 3/4	9 7/8
27	82	490	600	150	64	12 3/4	6 3/4	9 7/8
4D	165	1,050	1,340	355	135	20 3/4	8 1/2	10
8D	200	1,250	1,525	440	168	20 3/4	11	10

ADDING BATTERY CAPACITY

By increasing your battery capacity, you increase the amount of amp-hours that can be used before the battery is discharged to 50 percent of capacity. Also, the resultant battery bank can be charged with a higher charging current. The 50 percent of capacity of a 12 volt 200 amp-hour battery bank is now 100 amp-hours compared to only 50 amp-hours for a 12 volt 100 amp-hour battery. You can now safety charge the new battery bank with 50 amps (25% of 200 Ah). This resultant battery bank allows a longer period between charges and significantly reduces the charging time by accepting a higher charging current.

Two methods are used to increase battery capacity: connecting batteries in series and connecting batteries in parallel.

Batteries in Series

When it is necessary to increase the voltage, batteries are connected in series by having one battery's positive terminal connected to the negative terminal of another battery; see figure 4-1. Connecting two 6 volt 200 amp-hour batteries in series results in a 12 volt 200 amp-hour battery "bank."

Figure 4-1 Batteries Connected in Series

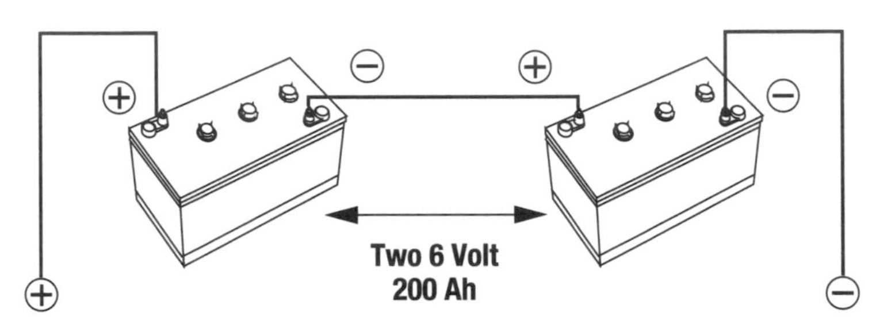

12 Volt 200 Ah Battery Bank

The resultant battery bank is now able to power the 12 volt loads on board your vehicle. The amp-hour capacity, however, is still the capacity of each battery.

This new battery bank no longer reacts electrically as two separate 6 volt 200 amp-hour batteries, but as one 12 volt 200 amp-hour battery. Just like the battery manufacturer who connects three 2 volt cells together in series to make the 6 volt battery, you can connect two 6 volt batteries in series to form one 12 volt battery bank. The battery bank powers 12 volt devices, and 12 volt charging systems are used to recharge this new battery bank—not a 6 volt charger.

Batteries in Parallel

Batteries are connected in parallel by having the positive terminals of two or more batteries connected, and by having their negative terminals connected; see figure 4-2. This increases the amp-hour capacity of the resultant battery bank, but not the voltage. By connecting two 12 volt 100 amp-hour batteries in parallel, you have a battery bank of 12 volts 200 amp-hours.

Figure 4-2 Batteries Connected in Parallel

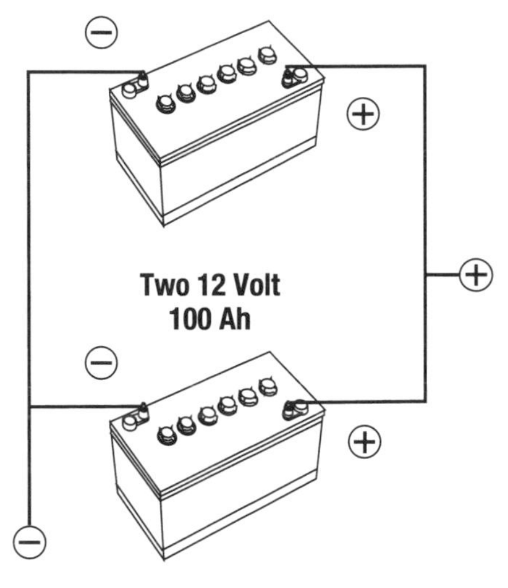

12 Volt 200 Ah Battery Bank

The Problem with Paralleling Batteries

The easiest method to increase capacity is to parallel a new 12 volt battery with an existing battery already installed on the vehicle. Or you could replace your existing battery with two new 12 volt batteries and connect them in parallel to increase capacity, like the example in figure 4-2.

The problem with paralleling batteries is if the batteries are not exactly equal in construction or degree of sulfation on the plates, the battery internal resistances are different. The battery with a lower internal resistance supplies a greater amount of current to a load than the battery with a higher resistance.

Also, if one battery has a higher voltage than the other, a small circulating current develops between the two parallel batteries. The higher voltage battery or battery bank is recharging the battery bank with a lower state of charge. The two batteries react just like two water tanks, one being full and the other being half empty. Water seeks its own level, so the full tank flows into the half empty tank until each tank has an equal amount. The same thing occurs with the two batteries, so you no longer have one fully charged battery, but two partially charged batteries. Since this process is inefficient, the battery self-discharge rates are increased and the battery bank's state of charge is reduced.

Paralleling an old battery with a new one results in the weaker battery determining the overall capacity of the total battery bank. A defective cell in one of the batteries does not support its share of the total load on discharge. On charge, it consumes a greater proportion of the current available to the detriment of the other batteries in parallel with it.

Even batteries of equal size from the same manufacturer and installed at the same time will develop cells weaker than the others, so that over time a small circulating current develops.

When a high capacity battery bank is needed, it is preferable to use two high capacity 6 volt batteries in series, rather than smaller capacity 12 volt batteries in parallel.

If your vehicle has the room and can withstand the bulk and weight of a large, high amp-hour 8D battery, it is better to have one battery rated at 200 amp-hours than have two smaller 100 amp-hours batteries in parallel.

In some situations, paralleling batteries is the easiest and simplest method; if so, the batteries should be matched as closely as possible. Don't mix maintenance free or gel cell batteries with normal wet lead acid batteries because they can accept different charges. It is best to use batteries from the same manufacturer.

Under no circumstances should you parallel a new battery with a dying battery because you will just kill the new one.

The new battery discharges into the dying battery trying to charge it, and will just discharge itself.

Having the ability to parallel batteries, however, can be helpful when the starting battery fails to start an engine. First, try to start the engine using another battery. If it also fails to start the engine, try paralleling the two batteries. The paralleled batteries may be able to start the engine.

DETERMINE THE BATTERIES FOR YOUR SITUATION

The batteries installed in RVs and boats by the manufacturers are sometimes undersized and inadequate for the vehicle's electrical devices. With the addition of inverters, color DC TVs, and other power hungry devices, the vehicle's battery capacity is usually inadequate for the demand.

Upgrading your batteries will improve your 12 volt electrical system more than any other single change.

The battery, or combination of batteries, best suited for your needs depends on your daily requirement. What also may be considered is the battery size, weight, and initial expense.

For example, if your daily requirement is less that 30 amp-hours per day then a small deep-cycle 12 volt 100 amp-hour battery is all you need. If your house battery is an automotive starting battery, the battery will soon be destroyed by deeply discharging it. Replacing it with a higher quality battery would significantly improve your electrical system.

If your daily requirement is less than 73 amp-hours, you will have enough power if you install two 6 volt 220 amp-hour batteries in series. These two batteries will support a daily requirement of 73 amp-hours per day (220 Ah divided by 3 = 73 Ah).

If your amp-hour requirement is higher, you can install a pair of 6 volt batteries with 350 amp-hours of capacity or install two banks of 6 volt batteries with 220 amp-hours each, and install a battery selector switch. Then you could switch between the two banks when one gets low. (Battery selector switches are explained in Chapter 6.)

Large high capacity 12 volt batteries, 4D and 8D, can also supply your daily requirement. They range in capacity from 165 to 225 amp-hours. For greater capacity, 2 volt single cells are available from 320 amp-hour to 1000 amp-hour. Six 2 volt cells can be connected in series to make a very large 12 volt battery bank.

Many different combinations could be used. Match your daily requirement to a combination of batteries and see which one makes the most sense.

Many large RVs and boats have large, high capacity batteries installed. Their battery capacity may be greater than three times the daily requirement, and yet the batteries are not keeping a charge or providing power for the vehicle. The problem, in this case, is not inadequate battery capacity but an inadequate charging system. Charging systems are discussed in the next chapter.

BATTERY MAINTENANCE AND SAFETY PRECAUTIONS

Batteries will provide years of service if properly maintained, but they can also be hazardous, so certain safety precautions must be observed.

- **When charging batteries, insure the battery area is vented.**
  Insure that no flames or sparks are present during charging because the potentially dangerous mixture of hydrogen and oxygen can explode. Do not have sensitive electrical equipment in the same compartment as a wet cell battery because the gassing of the electrolyte corrodes the electrical fittings. Also, the electrical equipment could supply a spark with potentially explosive results. Do not smoke near the batteries.

- **Sulfuric acid in the electrolyte causes skin burns and blindness if splashed in the eyes.**
  Wear eye protection. Wash all exposed areas quickly with water. Protect your clothing from the acid because the acid burns holes in clothing.

- **Rings, screw drivers, watches, and wrenches can melt if shorted across battery terminals.**
  A battery can produce hundreds of amps when shorted—enough current to melt a wrench. Remove all jewelry before working around batteries—fingers have been lost when rings melt. Insure that objects like screws, nails, and wrenches, cannot fall onto the battery terminals.

- **Salt water mixed with the electrolyte produces chlorine gas—a deadly mixture.**

- **Keep all battery tops clean and dry because dirty batteries self-discharge faster.**
  Water and dirt on the top of the battery promote corrosion at the terminals and may produce a path for electrical leaks. If any acid is present on the battery top, or corrosion is present on the terminals, clean and wash the battery using a solution of bicarbonate of soda. Coat the terminals with petroleum jelly to prevent corrosion.

- **Insure that all connections are tight.**
  Loose connections not only cause electrical problems but can cause explosions and serious injury if a spark occurs while the battery is gassing.

- **Never lift a battery by the terminals because internal damage could result.**

- **Do not allow the electrolyte level to drop below the top of the separators since this will shorten the battery life.**
  If the battery is using an excessive amount of water, it can be an indication of overcharging, high temperature operation and/or the end of the battery life. Too little water usage means undercharging and possible sulfation. After 30 to 50 hours of charging, the water loss should not be more than 2 fluid ounces per cell.

- **Use only distilled water when topping off the cells.**
  Tap water may contain iron and chlorine. Iron produces discharge on both the negative and positive plates, increasing self-discharge. Chlorine oxidizes at the positive plates during charge and eventually corrodes the plates.

- **Distilled water should be added to about 1/2 inch above the separators or to the fill level indicator.**
  Do not overfill.

- **Never add acid or additives to the battery.**
  Additional acid increases the specific gravity and increases corrosion of the plates. Battery additives have not been proven to have a positive effect on batteries and may shorten the battery life.

BATTERY STORAGE

When the vehicle is stored for more than a few months, the following procedure needs to be performed to keep the batteries healthy.

1. Clean the battery top and insure that the electrolyte is topped off with distilled water.

2. Bring the batteries to full charge.

3. Equalize the batteries. Perform this procedure (explained in Chapter 5) if you have a charger capable of performing an equalization charge.

4. Remove the batteries from the vehicle and store in a cool dry place or remove the battery cables from the battery terminals. The batteries can remain in the vehicle if the vehicle is attached to a battery charger or solar panel with float regulation. Insure all electrical loads are turned off.

5. Batteries not connected to a charging device need to be brought to full charge once per month. A trickle charger can be used to keep the batteries in good condition. Do not, however, leave a non-automatic trickle charger permanently attached to the batteries.

BATTERY INSTALLATION

Installing Batteries in the Vehicle

1. Batteries should be fully charged before installation. Check the voltage and specific gravity of each cell. The voltage should be 6.3 volts for a 6 volt battery and 12.6 volts for a 12 volt battery. Each cell's specific gravity should be 1.250 or higher and the difference between each cell should not be greater than 0.050. If the readings are lower, the battery will not perform properly.

2. Insure that the electrolyte level is correct, the battery top is clean, and the terminals are free of corrosion.

3. The battery must rest level in the battery container. Insure that no foreign object such as a stone or screw is under the battery. Such an object can wear on the case causing it to fail and result in the loss of acid.

4. Insure that the straps holding down the battery prevent battery movement and vibration. The holddowns should not be drawn tight, causing the case to distort or to crack.

5. Correctly install the battery cables. Reversing the positive and negative cables can cause serious damage to the electrical system. The negative or ground cable should be connected last.

6. Never hammer the cables on to the battery terminals. Damage to the cover, terminal connections, or plates could result. Insure that the cables make full contact to the terminal and are tight.

7. After the batteries are installed, operate the electrical system to insure that everything is working properly.

Chapter 5

Battery Charging

If you fail to charge a battery properly, the battery will fail you.

The house batteries on an RV or a boat have to support the electrical requirements of the vehicle. If they do not, people sometimes add more batteries, as much as 5 to 6 times their daily requirement, hoping the additional batteries solve their problems. The additional batteries do solve their electrical problem—at least temporarily. But, like a large fuel tank that enables you to travel long distances before having to refuel, the batteries eventually become empty. If, while dry camping or at anchor, your charging system can not adequately recharge the batteries, the batteries will fail no matter how large your battery capacity.

Once you have properly sized your battery capacity, the next question is: "How are you going to recharge your batteries adequately while dry camping or at anchor?"

The starting battery found on all engine driven vehicles, is used only to start the engine, so it only requires a small amount of energy to refill it. The vehicle's designer was more concerned about overcharging the starting battery than charging it quickly, so the designer purposely built the charging system to charge the starting battery at a low rate.

RVs and boats have installed large capacity batteries to power the 12 volt electrical loads when away from a 120 volt AC power source. But will the charging system, designed to prevent overcharging of a starting battery, produce enough energy to charge quickly those large capacity batteries? Just as a large fuel tank will eventually be filled using a low output hose, a low output charging system will also eventually charge large capacity batteries if it is run long enough. But who wants to run an engine for a long time to charge batteries while dry camping or at anchor?

Do you know the output of your alternator or battery charger? You now know your daily electrical requirement and your battery capacity, but how long do you need to run your charging system to replenish those lost amp-hours? Is the alternator or battery charger installed on your vehicle the most efficient way to charge your batteries in a relatively short time? Will solar power or wind power help to charge your batteries?

This chapter is devoted to solving the problem of recharging your batteries quickly and efficiently.

CHARGING DEVICES

Many different types of charging devices can be used to recharge batteries. Alternators are the most common charging device, but battery chargers and solar panels are also found on RVs and boats. Each device can be used as a stand alone charging device or can be used in combination with others to charge your batteries. It is important to understand the strengths and weakness of each device, so you can make an intelligent decision on which one is best suited for your situation.

ALTERNATORS

Alternators are a cheap and reliable way to charge batteries. Alternators are mass produced; every car, truck, and most boats have one. They have amperage ratings from 35 to 200 amps and produce these high rates at low cost compared to other charging devices.

The alternator generates electricity by rotating a magnetic field (rotor) within a circle of stationary, copper wire windings (stator). The amperage generated is determined by the strength of the magnetic field, the speed of rotation, and the number of turns of wire in the stator. The alternator output is controlled by adjusting the current (called the field current) to the magnetic field. A field current of about 1 amp to the spinning rotor induces 30 to 50 amps of output current on the stator. The voltage regulator controls the field current to the rotor. It is designed to keep the electrical system at a set voltage. Even a high amperage alternator will not charge a deeply discharged battery quickly unless the voltage regulator allows it. For this reason, it is important to understand voltage regulators.

REGULATING VOLTAGE

The function of the voltage regulator is to keep the electrical system at a set voltage level. It does this by adjusting the field current in response to the voltage of the battery and of the electrical system. The voltage drops when the battery is discharged or a load is placed on the system. The regulator responds to the voltage drop by increasing the alternator output, which increases the system voltage. As the voltage level is restored, the regulator tapers or reduces the current output.

The voltage regulator is temperature compensated to provide a slightly higher voltage at low temperatures and a lower voltage at higher temperatures to compensate for the charging requirements of the battery under these conditions.

A voltage regulator is similar to a pressure sensitive switch controlling a pump on a water tank. The pump is off when the tank is full and at a set pressure. The pressure in the tank decreases when a water valve is opened and the water flows from the tank. The pressure sensitive switch senses the drop in pressure and turns on the pump. When the tank is again filled with water and the pressure reaches the set point, the pressure switch turns off the pump. In the electrical circuit, a light switch (water valve) turns on a light that reduces the voltage at the battery (water tank). The voltage regulator (pressure sensitive switch) senses the voltage (pressure) drop and increases the alternator's (water pump) flow of electrons (water) in the electrical system. As the voltage increases and nears the set voltage, the flow of electrons or current is reduced.

The voltage regulator's control over the alternator output can be observed by watching the ammeter while a standard automotive alternator is charging. The starting battery voltage will drop significantly while starting the engine. The voltage regulator senses the low voltage and increases the field current to the alternator, resulting in higher alternator output. As the battery voltage nears the set voltage of the regulator, the field current is decreased and the alternator output current decreases. In this example, the ammeter reading initially will be high but soon drops considerably.

The voltage regulator also reacts to any voltage drop due to electrical loads being turned on. If you turn on the headlights of a car and draw an extra ten amps, the regulator increases the current from the alternator—it takes the additional electrical load instead of the starting battery. Both of these functions work to solve common problems associated with an automobile where the battery is only used to start the engine, but this type of regulator does little to solve the problems RVers or boaters have with deeply discharged batteries.

AUTOMOBILE REGULATORS

Automobile starting batteries are used just to start the engine and are rarely discharged by more than 5 percent of their capacity. The alternator, not the battery, provides the energy for the vehicle electrical loads while the engine is running.

Since the alternator is continuously producing current, there is a concern that the starting battery will be overcharged and outgas hydrogen and oxygen. To prevent overcharging, a regulator is needed that will supply a large amount of current initially to bring the battery voltage to a high level, but once the voltage is relatively high the current is drastically reduced. The **constant voltage** (potential) method of regulation solves this problem. A constant voltage regulator seeks to maintain output at a set voltage. If the battery voltage decreases, the charger responds by increasing output. As the battery voltage increases, the output is decreased or tapered.

Figure 5-1 Constant Voltage or Taper Chargers

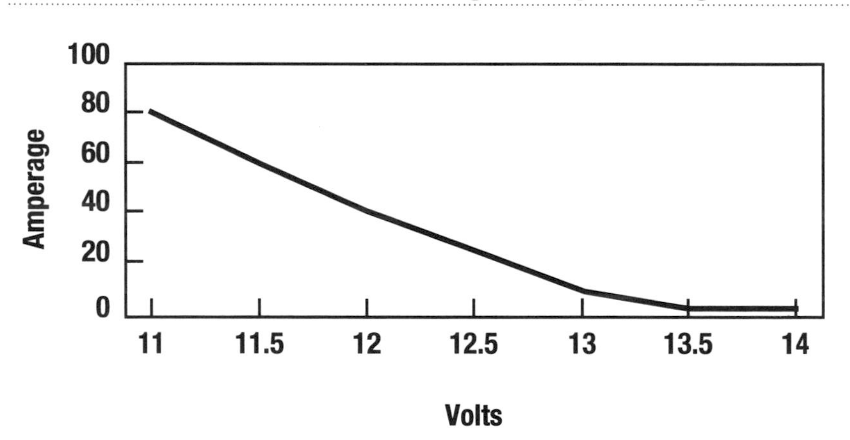

Figure 5-1 illustrates what happens when a 75 amp alternator with constant voltage regulation charges a deeply discharged battery of 11 volts. The voltage regulator is set to regulate the voltage at 14 volts.

Initially the alternator produces 75 amps, but the output is quickly decreased as the battery voltage increases. At 13 volts, the 75 amp alternator output is only 10 amps, and at 14 volts, it produces 2 amps. Batteries should not be discharged to 11 volts, so 75 amps are rarely produced from the alternator. If the battery had only been discharged to 12.5 volts, the constant voltage regulator would initially produce 26 amps, decreasing rapidly as the voltage increases.

Constant voltage chargers are used mainly on systems where the batteries will not be discharged to greater than 25 percent of capacity. These systems are quite safe. They prevent the charging system, when operated over a long period, from overcharging the batteries.

THE PROBLEM WITH STANDARD AUTOMOBILE VOLTAGE REGULATORS

In figure 5-2, the left side of the graph shows how battery voltage and state of charge, as measured in amp-hours, are effected during discharge. The voltage decreases steadily until the final voltage of 10.5 volts is reached. As the battery voltage decreases, the battery's state of charge decreases linearly.

Figure 5-2 Battery Voltage and State of Charge

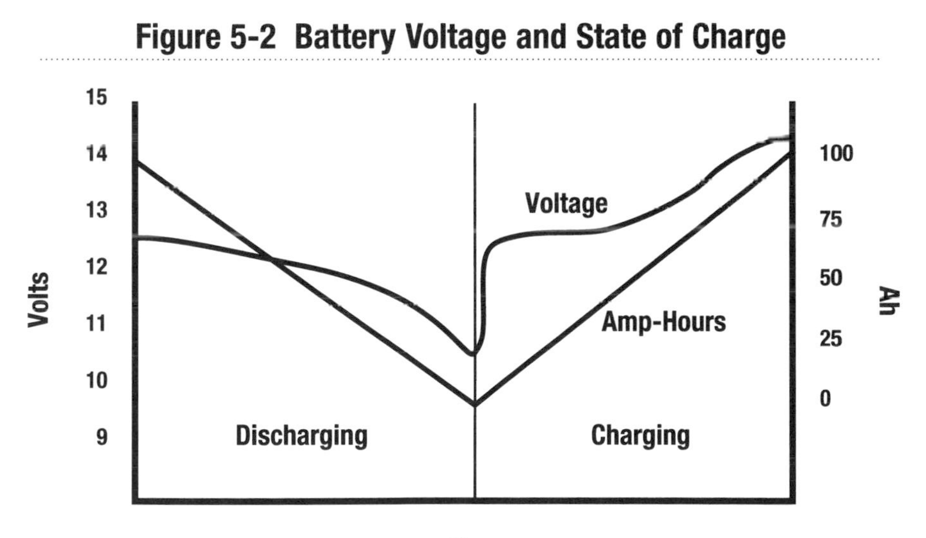

The right side of the graph shows that when the battery is first put on charge, a sharp rise in the battery terminal voltage occurs. This is probably due to a sudden increase in concentration of the electrolyte near the plates as the sulfuric acid is driven back into the electrolyte. Even though the voltage increases dramatically, the battery's state of charge, as measured in amp-hours, does not increase as rapidly as the voltage increases. It takes time for the sulfuric acid to diffuse to the outer reaches of the cell. During charging, the voltage does not indicate a battery's state of charge.

A standard automobile voltage regulator using constant voltage regulation senses the increase in voltage and reduces the alternator output. So fewer amps are being returned to the battery, increasing the time it takes to recharge the battery. To restore the battery's state of charge, amperage must be returned to the battery over a period of time. If you charge a battery for one hour with 20 amps, 20 amp-hours is returned to the battery. But if you charge a battery with only 5 amps, it takes 4 hours to return 20 amp-hours to the battery. (Plus an additional 20 percent must be charged because of battery inefficiency.) The higher the current the less time it takes to recharge the battery. With the decrease in amperage when using constant voltage regulation, it takes a longer time to recharge a deeply discharged battery. This is not a problem when time is of little concern. But if you want to charge a deeply discharged battery quickly, constant voltage regulation is not the answer.

Automobile constant voltage regulators solve the problem of preventing alternators from overcharging a starting battery, but will not charge a deeply discharged battery quickly.

MULTI-STAGE CHARGING

You learned earlier that a deeply discharged battery accepts a large amount of current until the gassing voltage is reached. Since a constant voltage regulator reduces the amperage well before the gassing voltage is reached, another type of regulation is needed for the first step of charging a deeply discharged battery.

A **constant current (amperage)** regulated charger produces a fixed amount of current. If the charging current is at a high setting for a deeply discharged battery, the constant current will charge the battery quickly. As the battery approaches full charge, however, the amperage must be reduced because battery damage could occur.

Incorporating the best features of constant current and constant voltage charging solves the problem of quickly and efficiently charging a deeply discharged battery.

Sophisticated multi-stage regulators and battery chargers have been developed using the constant current/constant voltage charging method to charge deeply discharged batteries quickly and efficiently. This charging method is in three distinct stages: bulk, absorption, and float. Some chargers have a fourth cycle called the equalization cycle, which should be periodically utilized to insure that deep cycle batteries have a long and useful life.

Bulk Charging Stage

During the bulk stage, the multi-stage charger produces a constant current; the amperage is kept at a high rate until the gassing voltage is reached. Therefore, more amperage is absorbed by the battery, reducing the charging time. This is unlike a constant voltage charger, which tapers the charge, reducing the amperage as the voltage increases.

Figure 5-3 Bulk Charge Versus Taper Charge

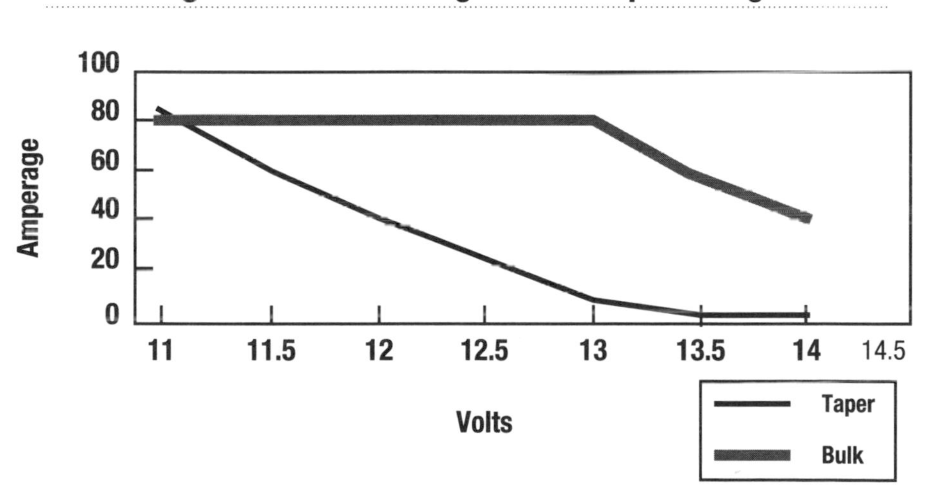

Figure 5-3 compares the bulk charging stage to a taper charge. During the bulk charge, the amperage is held at a high rate even as the voltage increases (but decreases slightly as the voltage nears 14 volts). The taper charge is reduced as the voltage is increased. Since the amperage is maintained at a high level during the bulk charge, the battery's state of charge is restored more quickly.

The optimum rate of charge for the bulk stage is 20 to 25 percent of the battery capacity. As Table 3-4 on page 39 indicates, a rate of charge of about 25 percent drastically reduces the charging time compared to a charging rate of 10 percent of capacity. The battery's state of charge at the gassing voltage is not that much less than it would be if a rate of 10 percent were used. In the example of a 100 amp-hour battery, a current of 25 amps can be used to recharge the battery. A 200 amp-hour battery bank can accept a current of 50 amps during the bulk stage.

Absorption Stage

Even when the gassing voltage is reached, the battery is not fully charged. Its state of charge is only 60 to 80 percent of capacity depending on the charging rate. The battery cannot accept a large amperage at the gassing voltage, so the amperage must be reduced or the excess amperage will decompose the electrolyte into its component elements of hydrogen and oxygen, and generate excessive heat.

When the voltage reaches approximately 14.4 volts, the multi-stage charger trips from a constant current charger to a constant voltage charger. This is called the absorption stage. The amperage is reduced to the natural absorption rate of the battery, but the voltage is maintained at about 14.4 volts. Any charging current that is greater than the natural absorption rate only serves to heat the battery and reduce its life. Some battery manufacturers recommend that the absorption rate be set at 13.8 volts, the same voltage that standard voltage regulators and battery chargers are set to. At this level, the problem of gassing is minimized, but it takes longer to increase the battery state of charge to 90 percent. It is better to keep the voltage at 14.4 volts until the battery acceptance rate decreases to less than 10 amps. Once the absorption rate drops to a low level, the voltage should be reduced to around 13 volts, so the multi-stage charger switches to a third stage. If the voltage remains high, the battery can be overcharged, causing corrosion of the positive plates and water loss. Because the battery absorption rate is low for the last 20 percent of capacity, it takes many more hours to reach 100 percent state of charge.

Float Stage

Once the battery approaches full charge and the absorption rate drops to a low level, the multi-stage charge trips to a voltage less than the gassing voltage. This is called the float stage, and it maintains the battery voltage at about 13 volts depending on temperature: the higher the temperature the lower the float voltage. When the battery voltage drops below this level, the charging device provides enough current to bring the battery back to this level. The battery "floats" at this voltage. The float stage prevents self discharging, and the charger supplies the current needed when a load is placed on the system. Corrosion of the positive plates and water loss due to overcharging are kept to a minimum. The battery should be removed from the float stage and discharged and recharged at least every two to three months. A battery needs to be periodically "exercised" to keep it in full form.

Equalization Stage

During normal charging, not all the sulfate is removed from the plates. Over time, the sulfate hardens, decreasing the area where the electrochemical process takes place, which reduces the battery capacity. This process is called sulfation. Initially the sulfate is relatively soft and porous, but if the battery is left in a discharged state, the sulfate hardens, reducing the capacity of the battery.

The soft sulfate can be reconverted by bringing the battery to full charge using a controlled extended charge. This process is called equalization or conditioning and is especially important for deep cycle batteries that are not always brought back to full charge on a regular basis. This procedure may be required as often as once a month.

Removing the sulfate requires boosting the charging voltage to as high as 15 or 16 volts, but the current must be held to 3 to 5 percent of the battery amp-hour rating. If the amperage is not controlled, the internal battery temperature can exceed 125° F, and the battery can be cooked. Normal voltage regulators and battery chargers do not have the voltage or amperage control to perform equalization cycles. Only adjustable regulators and battery charges can be used for this procedure.

Some sophisticated battery chargers, voltage regulators, and inverters with multi-stage chargers, which can be installed on RVs and boats, can perform equalization cycles. Follow the manufacturer's instructions when using these chargers to equalize your batteries.

Equalize only one battery at a time, and insure that all electronics are turned off; some sensitive electrical equipment cannot tolerate voltages a high as 16 volts. If you are using an adjustable voltage regulator, slowly increase the output, so the voltage is greater than the gassing voltage, around 14.4 volts. Watch the ammeter to insure the amps do not increase above 5 percent of the rated capacity of the battery. Continue the current at the 5 percent level. The voltage may increase to as much as 15 to 16 volts. Cut back the amperage if the voltage is increasing rapidly.

Periodically, take a specific gravity reading using a hydrometer. The specific gravity reading for a fully charged battery is around 1.265, but it varies from battery to battery and with temperature. One or two cells may have a specific gravity reading lower than the rest. Continue with the equalization charge until all cells have reached the same high reading of about 1.26. If after several hours, one cell continues to be lower than the rest, by 0.050 or more, the cell may be dead. Once the weakest cell reaches full charge, the equalization process is complete. After conditioning your batteries, let them rest, and then measure the battery specific gravity and voltage. These measurements are the "full" mark for your batteries.

During equalization, gassing occurs, with the electrolyte bubbling and releasing hydrogen and oxygen. The battery compartment and the vehicle should be adequately vented, and insure that nothing is creating a spark. Also, wear some type of eye protection. After the conditioning cycle is complete, top off the cells with distilled water and wipe the area with a solution of bicarbonate of soda .

Never attempt an equalization charge on sealed or gel cell batteries. The sealed tops prevent water from being added or the gas pressure from being relieved.

Equalization has its hazards and may not be worth the effort for inexpensive batteries, however, for expensive deep-cycle batteries this procedure extends their life.

ALTERNATORS WITH MULTI-STAGE VOLTAGE REGULATORS

Several marine product companies have designed and are manufacturing sophisticated multi-stage voltage regulators that automate the charging process. These regulators follow the bulk, absorption, and float stages as described above. A multi-stage regulator coupled to a high-output alternator makes a very powerful charging system. Most of these systems are packaged with a high output alternator and a multi-stage regulator. The alternators range in size from 75 to 150 amps.

A multi-stage regulator can be installed with a vehicle's existing alternator. Unfortunately, normal automotive alternators were not designed to operate at their rated output for a prolonged period. A standard automobile alternator should not be operated above 50 to 65 percent of its rating. If the your battery capacity is 200 amp-hours, you can recharge them at 25 percent of their capacity during the bulk cycle, or 50 amps. At 50 amps, however, a standard 55 amp alternator is going to be destroyed very quickly. In this case, the alternator must also be upgraded to take advantage of the multi-stage voltage regulator. High output alternators are available. Emergency vehicles are equipped with high output alternators that provide 100 to 200 amps at 90 percent of their rated output.

BATTERY CHARGERS (CONVERTERS)

Converters perform the opposite job of inverters. Converters take the 120 volts AC coming from the electrical utility or generator and feed it into a transformer, a device for changing voltage. The transformer steps down the voltage to 12 volts compatible with the battery, but it is still alternating current (AC). The AC current must be rectified by silicon diodes or other solid state devices, like an SCR, to turn the current into direct current (DC). Now you have 12 volt DC that can power your lights, stereo, and charge your batteries.

Battery chargers and converters, like other electrical devices, are designed for specific purposes. The one that came with your vehicle may be well-suited for what the manufacturer wanted, but the battery charger may be ill-suited to charge deeply discharged batteries quickly and efficiently.

Portable Taper Chargers

A **trickle charger** is unregulated and can have an output from less than an amp to around 3 amps. A continuous charge is applied to the battery, regardless of the state of charge. If a large battery is deeply discharged, it will take hours, if not days, to bring the battery to full capacity using a trickle charger.

Boost battery chargers with outputs of 6 to 50 amps are available at auto parts stores; see figure 5-4. They have an automatic setting that controls the charging current; as the battery voltage increases, the charging current tapers to a low charging current. In the automatic setting, many of these chargers can be connected indefinitely to a battery because they cycle on and off as necessary. In the manual setting, the boost charger produces a larger amount of amperage that can rapidly recharge batteries but can overcharge them if not turned off when the battery starts to gas. Some boost chargers have a timer that switches off the charger after a preset duration; others need to be monitored. Boost chargers could be used to charge your batteries if you monitor the voltage to insure that it does not climb higher than the 14.4 volt gassing voltage.

Portable generators feature a DC output of around 8 to 10 amps. If the daily electrical requirement is low, they can be useful in charging batteries. If the daily electrical requirement is high, a portable generator's DC output may not be sufficient to charge a high capacity battery, because it will take a long time.

Figure 5-4 Portable Battery Charger

Constant Voltage Chargers

Most battery chargers are constant voltage chargers that regulate their output similarly to the voltage regulator on an automobile. They seek to maintain output at a constant voltage. These chargers have the same problem as standard automobile regulators: they do not charge a deeply discharged battery quickly.

Most converters/battery chargers are ferro-resonant transformers, a type of constant voltage charger. These chargers reduce manufacturing costs because the transformer automatically stabilizes output voltage at a certain level, which eliminates the need for various control circuits.

RVs are equipped with converters that have an output of 25 to 50 amps depending on the model. A battery charger is usually associated with the converter, but it is rated at only 3 amps, about the output of a trickle charger.

The converter and not the battery supplies the power to the 12 volt lights, pumps, and fans of the RV. When the 120 volt AC power is shut off, an automatic relay switches the load to the battery.

A separate unit of this converter contains a battery charger. It "senses" the condition of the battery and charges it as necessary, but its output is little more than a trickle charger, about 3 to 6 amps. At this rate, it takes over 5 hours of generator time to replace a daily electrical requirement of 30 amp-hours. Plus an extra 20 percent must be recharged due to the inefficiency of the charging process. RV manufacturers often install a 3 amp charger to cut costs and to keep the amperage low, so the batteries will not be overcharged while you're plugged into an AC outlet at a campground. This battery charger will not recharge your batteries quickly.

An RV and a marine constant voltage charger's output may have a rating as high as 70 amps. However, what is not listed is the voltage at which the charger produces 70 amps. Some chargers produce 70 amps at 11 volts, which is a dead battery. When the battery reaches 12.2 volts, this 70 amp battery charger may only produce 35 amps. At a voltage of 13 volts, the battery charger is only producing a trickle charge. A constant voltage charger tapers the amperage quickly as the voltage increases, so the rated amperage is only produced for a very short period.

When replacing a battery charger, it is important to know at what voltage the charger supplies its rated output. Ask what amperage the charger produces at 13.8 volts. If the charger output is little more that a trickle, you may as well keep your ferro-resonant charger and keep looking for a better one.

RVs are sold with the belief that the generator and battery charger installed on the vehicle will quickly replace the amp-hours used each day. Unfortunately, the standard battery chargers found on RVs and boats are constant voltage chargers and are not designed to charge batteries quickly.

Multi-stage Battery Chargers

Three stage or multi-stage battery chargers have been developed to charge deeply discharged batteries quickly and safely. Some of them use microprocessors to control the multi-stage constant current/constant voltage charging process. These battery chargers follow the bulk, absorption, and float stages described above.

This type of charger is ideal for the vehicle with a generator. Even a 650 watt portable generator can power a 20 amp multi-stage battery charger if no other 120 AC loads are on. A sophisticated multi-stage battery charger, powered by a generator, can recharge batteries relatively quickly, allowing the batteries to support all the electrical loads found on modern RVs and boats today.

Inverters with Multi-stage Battery Chargers

Inverters are purchased mainly for their ability to transform battery voltage (12 volts DC) to household voltage (120 volts AC) so that normal household appliances can be used without turning on a generator or being connected to an electrical utility. But the more powerful and expensive inverter models also have incorporated multi-stage battery chargers that quickly and safely recharge your batteries. When the inverter is connected to a 120 volt AC electrical utility or to a generator, the inverter's multi-stage battery charger recharges batteries in the same manner as the constant current/constant voltage battery chargers described above. The battery chargers incorporated into the inverters have outputs of 25 to 130 amps depending on model. The units are expensive, but they do an excellent job of providing AC power when not connected to 120 volts AC and of recharging batteries when connected to 120 volts AC.

Table 5-1 Battery Charger Specifications

Types	Amperage output	Dimensions	Cost
Portable taper charger with multi-setting	10 amp, 30 amp boost, 50 amp engine start	11" x 5" x 9"	$40-$70
multi-stage charger	20	15" x 6" x 2.25"	$300-$400
multi-stage charger	40	15" x 6.5" x 2.5"	$350-$500
750 watt inverter & multi-stage charger	25	12" x 10" x 7"	$500-$750
1000 watt inverter & multi-stage charger	50	12" x 10" x 7"	$700-$1090
2000 watt inverter & multi-stage charge	100	12" x 10" x 9"	$1100-$1650
2500 watt inverter & multi-stage charger	130	12" x 10" x 9"	$1300-$1600

SOLAR PANELS

Solar panels convert sunlight into electrical energy by use of photo voltaic cells. Each of these cells, which are essentially silicon based semiconductors, generates about 0.45 to 0.5 volts in ideal direct sunlight. To charge a 12 volt battery, a panel must have at least 32 cells in series to produce a voltage high enough. A panel of 32 cells produces 14 to 16 volts and about 3 amps or 48 watts at 120°F. Obviously, the amount of sunlight that strikes the panel determines how effective the solar panel will be. On a cloudy day, the output will be significantly less.

Several factors determine how effective the panel will be on a sunny day. The solar panel voltage increases as soon as light strikes the panel, but direct sunlight is necessary before the solar panel produces a significant amount of current or amperage. The solar panel must be at a 90° angle to the sun to obtain the maximum current and must be constantly turned as the sun passes overhead. The optimal sunlight striking the earth is around 9:30 am to about 2:30 pm, so the panel's rated output is only achieved for about 5 hours a day. Temperature of the panel affects the voltage output—the higher the temperature the lower the voltage. At noon, the greatest current output occurs, but also the highest temperatures occur, so the voltage is at its lowest. To overcome this problem, a panel with as many as 36 cells is required. A 36 cell panel with a voltage as high as 18 volts, if left attached to a battery could cause damage. However, at 3 amps an hour on a 5 hour sunny day, the daily result is only about 15 amp-hours. If the daily requirement is 50 amp-hours, the panel is not supplying the energy the batteries need, nor is it harming the battery. If the battery is fully charged, the solar panel could cause damage. A regulator should be attached to a solar panel that is providing energy to a seldom used battery. Self-regulating solar panels advertised in the market today contain fewer cells, so the voltage is low. They do not produce enough voltage to harm the batteries, nor do they effectively charge a deeply discharged battery.

A solar panel connected directly to a battery discharges the battery at night, so a blocking diode is placed in line between the panel and the battery. The diode has a voltage drop of 0.6 volts and this voltage drop generates heat and looses energy. A solar panel with an output of 3 amps at 14 volts without a blocking diode will only produce 3 amps at 13.4 volts with a blocking diode. This is not enough voltage to recharge a battery efficiently.

With a daily requirement of 50 amp-hours, at least three 3 amp solar panels are required to replenish the batteries. Given that they only work on sunny days and must be turned constantly to be at a 90° angle to the sun, a solar panel is not the most practical, reliable, or cost effective way to replenish your daily electrical requirement. Under certain circumstances, however, solar panels can be effective battery chargers.

In an area where the sun constantly shines, solar panels can provide adequate power if the daily requirement is low or you have enough solar panels. Also, solar panels, as an element in a total charging plan, can charge batteries during the absorption stage after an alternator or a battery charger with multi-stage voltage regulation charges the batteries during the bulk stage.

During the float stage, solar panels are effective at preventing self-discharging when the vehicle is left unattended for a long time. The self-discharge rate of wet lead acid batteries is around 0.5 percent per day at 80°F and 1 percent per day at 100°F. Older batteries could have a self-discharge rate at twice this rate. A 100 amp-hour battery could lose as much as 0.5 to 1 amp-hour a day. To help prevent batteries from self discharging, a 5 watt solar panel with an output of about 1.5 amp-hours a day could be installed for every 200 amp-hours of battery capacity. An unregulated solar panel could safely be attached to a battery bank if the panel output is less than 1 percent of the bank's rated capacity.

WIND GENERATORS

Cruising sailboats can use the free power of the wind as a power source for charging batteries. Like solar panels that need a sunny day to operate, wind generators only function properly in a steady wind. In areas of the world where the wind fluctuates between calm and strong weather fronts, the wind generator output is extremely erratic. In the tradewinds, however, wind generators can be a reliable power source.

Wind generators have come a long way from the time when people experimented with an aircraft propeller attached to an alternator. They hung the wind generators in the boat rigging with numerous lines to keep them from hitting the mast and rigging. The wind generators were so noisy you always felt you were anchored next to the airport when the wind picked up. Now small, relatively quiet units are available that can be permanently mounted to the boat, which permits them to work even under sail.

The two types of wind generators are alternators and permanent magnet DC motors. Alternator type wind generators operate similarly to your engine alternator, and require high rpms to operate efficiently. The permanent magnet DC motor, the most efficient form of wind generators, reverses the alternator mechanism by spinning a series of coils inside a set of fixed magnets. This DC motor will produce about 15 amps, compared to an alternator that only produces 8 amps, given the same amount of wind.

Like other charging devices, wind generators must be regulated. A voltage regulator is necessary to control the voltage going to the batteries. Some regulators have an upper limit cutout that automatically breaks the charging circuit when the battery voltage reaches a certain preset level. Another method switches the wind generator output to a "dummy" load, or shunt. The shunt regulator is a solid state device that diverts the current from the batteries to a heat sink. When the voltage approaches a preset level of 13.8 volts, for example, the shunt regulator increasingly diverts the wind generator output from the battery to the heat sink.

A diode must be placed in the generator wiring line to prevent the batteries from discharging to the wind generator when the wind dies.

In high winds, the wind generator can become very dangerous. DC motor types have unlimited output as their speed increases, and it will burn out in high winds under no load. Some can be "braked" electrically; others must be turned 90 degrees into the wind and stalled before they can be shut down.

In the tradewind areas of the world, a wind generator can provide ample energy to recharge a cruising sailboat's batteries.

DETERMINE CHARGER OUTPUT BASED ON YOUR BATTERY CAPACITY

A multi-stage charger, with an output of 25 percent of your battery capacity, is the most effective device to charge your batteries quickly and efficiently.

Your battery amp-hour capacity ______ x 0.25 = ______ output rating of a charging device.

Many choices of charging devices and combination of devices are available to recharge your batteries. The most reliable method is a multi-stage voltage regulator controlling your alternator, or a generator powering a multi-stage battery charger, but other methods have their place in a successful charging plan. If solar panels are installed, they can reduce the time the engine or generator has to operate. A cruising boat in the tradewinds may provide most of its daily requirement with a wind generator. A small RV with a low daily requirement may just need a portable generator powering a portable battery charger to replace its daily consumption. No one answer is suitable for everyone; just what works for you.

You now have the answer to the third question asked in Chapter 1, What type of charging device or devices do you need and how much output should be produced to recharge your battery quickly and efficiently?

If you still have questions on what charging device is appropriate for your situation and how it would work on your vehicle, recommendations are made in Chapter 7 for four different vehicles and situations.

Chapter 6

Monitors, Wiring, and Switches

***If you can monitor what is occurring,
you can respond to problems.***

If you do not have a fuel gauge on your vehicle, how do you know how much fuel is in your tank? How do you know when you are running low on fuel? You can only guess and if you guess wrong, you will run out of gas.

How do you know how much "fuel" is in your batteries? How long will they provide power? Are you just guessing that they will support your electrical requirements? If you had electrical monitors you could answer these questions and you could successfully manage your 12 volt electrical system.

Determining how "full" your batteries are or their state of charge is not straight forward. There is no simple "fuel gauge" for your batteries. A voltmeter will indicate a battery's state of charge, but the battery must be at rest and at 80°F for the reading to be accurate. This is not always possible.

Understanding a battery state of charge and the rate of charge and discharge is important. The addition of a couple of meters greatly enhances your ability to control your 12 volt electrical system.

The two most important electrical meters are:

- **The voltmeter, which measures the voltage or pressure of the electrical system;**
- **The ammeter, which measures current or rate of flow.**

VOLTMETERS

A voltmeter measures the battery or the system voltage. If you know the voltage, it will provide a clue as to what is occurring with the 12 volt electrical system or the battery. Since only one volt separates a fully charged battery from a "dead" battery, a digital voltmeter is more accurate than an inexpensive analog voltmeter.

Voltage Readings

1. Battery voltage at rest

- Battery voltage is an indicator of its state of charge if the battery has not been discharged or charged for at least 15 minutes and preferably longer—four hours or overnight. At temperatures above 80°F, the voltage will be higher, and at lower temperatures, the voltage is also lower. Table 6-1 indicates a battery's state of charge as it relates to its voltage.

Table 6-1 Battery Voltage and State of Charge

Open Circuit Voltage	State of charge
12.6 or greater	100%
12.4 to 12.6	75% to 100%
12.2 to 12.4	50% to 75%
12.0 to 12.2	25% to 50%
11.7 to 12.0	0 to 25%
11.7 or less	dead

- A good battery properly charged has a voltage reading above 12.6 volts.

- A voltage reading higher than 12.6 volts indicates a "surface charge" on the plates. Place a load on the battery to remove the surface charge and read the battery voltage again.

- A voltage reading below 11.7 volts indicates a "dead" battery caused by undercharging, a short in the electrical system, or a worn out battery.

2. Battery voltage under load

- When an electrical load is placed on a battery, the voltage decreases: the greater the amperage the greater the voltage drops. When the load is turned off, the battery voltage recovers or increases.

- A starting battery's voltage drops to 11 volts or less when supplying amperage to the starter motor, but recovers to above 12 volts after the engine starts. When the engine is running at normal speeds for a few minutes, the voltage increases to above 13 volts. If not, the battery is not being charged.

3. Battery voltage while charging

- The voltage should rise to 13 volts or higher when the alternator or battery charger is turned on. If not, a problem exists with the charging device, the battery, or the connections in the charging circuit.

- With all electrical loads turned off, the highest voltage obtained during charging indicates the voltage regulator set point. If the battery is not fully charged, the voltage is less than the charging device set point.

- The voltage, when being charged by a multi-stage charger, should climb to over 14 volts during the bulk and absorption stage.

- When being charged by an alternator, the battery voltage should increase to around 14 volts.

- A battery in the float stage should be at 13.1 to 13.9 volts depending on temperature: 13.1 volts at 90°F to 13.9 at 50°F for a wet cell battery. If the voltage is high for the temperature, the battery may be overcharged resulting in water loss and reduced battery life.

- A gel cell battery can be damaged if charged at voltages higher than 14.2 volts.

- Voltages above 15 volts indicate overcharging, which results in water loss and shortened battery life. Check for a defective voltage regulator or battery charger.

Where to Install a Voltmeter

A voltmeter is installed with its negative terminal connected to a wire going to ground or to the battery negative terminal; see figure 6-1. The voltmeter positive terminal is connected to a wire from the battery positive terminal or positive side of the circuit.

Figure 6-1 Voltmeter

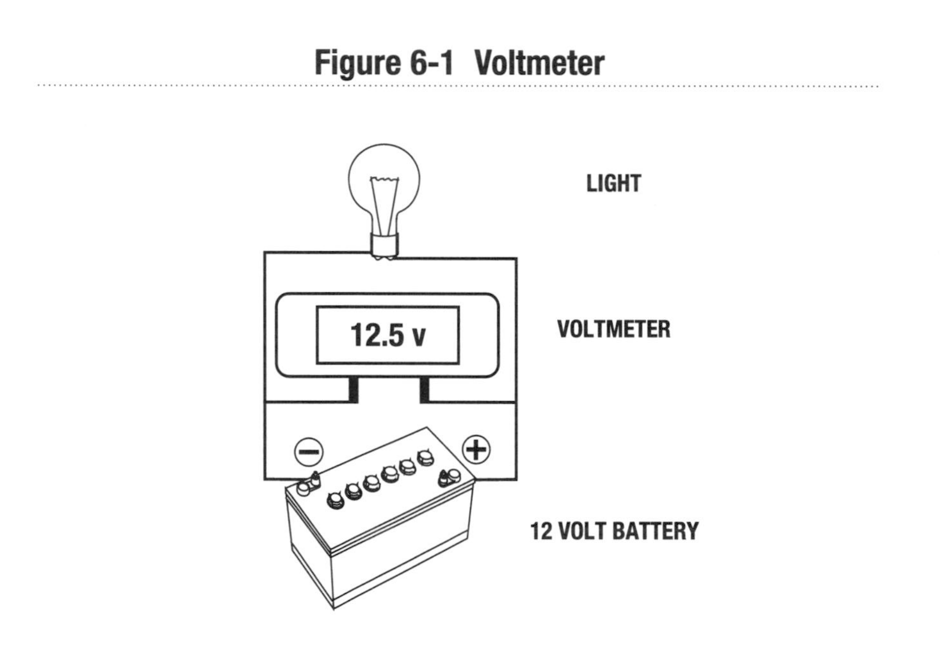

AMMETERS

Ammeters measure the electrical current in a circuit. They cannot indicate the batteries' state of charge, but do provide additional clues as to what is occurring in the electrical system.

Ammeter Applications

There are two common uses of an ammeter on an RV or a boat:

1. An ammeter can indicate the amount of current electrical loads draw.

- Conservation of your battery energy is extremely important. Mount this ammeter in a location where it can easily be observed. This allows you to respond to unusual rates of discharge, which if not corrected could completely discharge your batteries. For example, if you believe that all electrical devices are turned off, but the ammeter indicates a discharge rate of a few amps, something is still on. Upon investigation, a small light in the toilet or engine room may have been left on, and if this light had remained on, the battery's state of charge could have been depleted. **If an electrical device is not needed, turn it off.**

- An ammeter helps in determining the daily electrical requirement. An electrical device's amperage rating can be determined by turning on the device and observing the amperage increase on the ammeter.

2. An ammeter is used to measure the current output of a charging device.

- Is the charger producing enough amperage to recharge your batteries quickly? If the battery charger is only producing 3 amps, and your daily requirement is 50 amp-hours, it is going to take a long time to recharge your batteries.

- Observe the battery voltage along with the amperage the charger is producing. A taper or constant voltage charger's amperage is high only at low voltage. The amperage decreases as the voltage increases. A 10 amp taper charger produces only 6 amps at 13 volts and drops to a few amps at 13.8 volts.

- If the charging device is rated 20 amps at 14 volts, is this what you are getting or is it less? If you are using a multi-stage regulator, are the amperage and voltage readings following the bulk, absorption, and float cycle described in Chapter 5?

Where to Install Ammeters

Ammeters are placed in the electrical circuit to measure the current through a circuit; see figure 6-2.

Figure 6-2 Placement of Ammeters

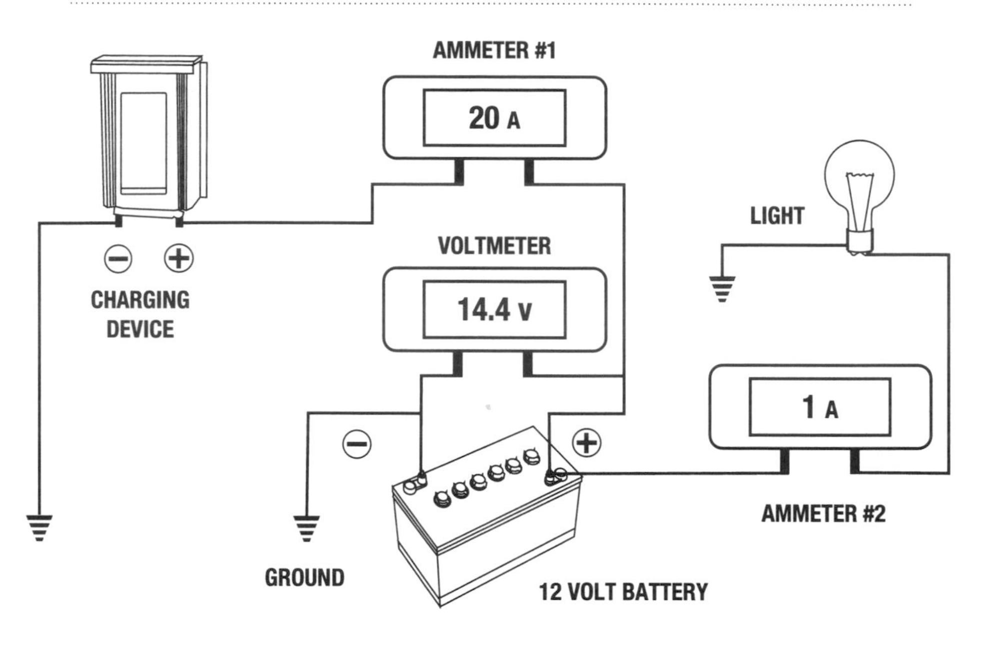

Ammeter #1 installed on the battery charger positive side shows that it is producing 20 amps. Ammeter #2 shows that the light, which is on, is consuming one amp. The other 19 amps are recharging the battery. Since the current is the same on the positive and the negative side of a circuit, an ammeter can be connected to the negative battery cable to measure the current the battery is providing to loads or receiving from a charging device.

The meter must be matched to the maximum amount of current the circuit will carry. Installing a 25 amp ammeter in a 100 amp circuit will not only destroy the meter but could cause a fire. An ammeter must be large enough to withstand the circuit amperage, or the ammeter will be destroyed.

In electrical systems where the amperage is very high, a special ammeter may be used. At high amperages, a shunt is placed in line to measure the current instead of placing the ammeter directly in the circuit. A shunt is a low resistance device whose resistance is known. As the current flows through the shunt, the voltage drops by a small amount and a millivolt meter measures the voltage drop to determine the actual current. If the shunt is rated at 1 millivolt (one thousandth of a volt) per amp and the voltage drops 5 millivolts, the current through the electrical system is 5 amps.

ELECTRICAL PANELS

Sophisticated electrical panels have been developed with digital displays that can monitor up to 4 banks of batteries. These units have incorporated voltmeters and ammeters into one compact package. By turning a knob, or flipping a switch, you have either the amperage or voltage reading for any of your batteries. Knowing the discharge or charging rate will never be a mystery again. The panel's compact size is preferable to having to install a voltmeter and one or two ammeters.

Figure 6-3 shows an electrical panel connected to two batteries or battery banks. With a turn of a knob, the panel can display either battery's voltage or amperage that is being supplied to a load.

Figure 6-3 Electrical Panel

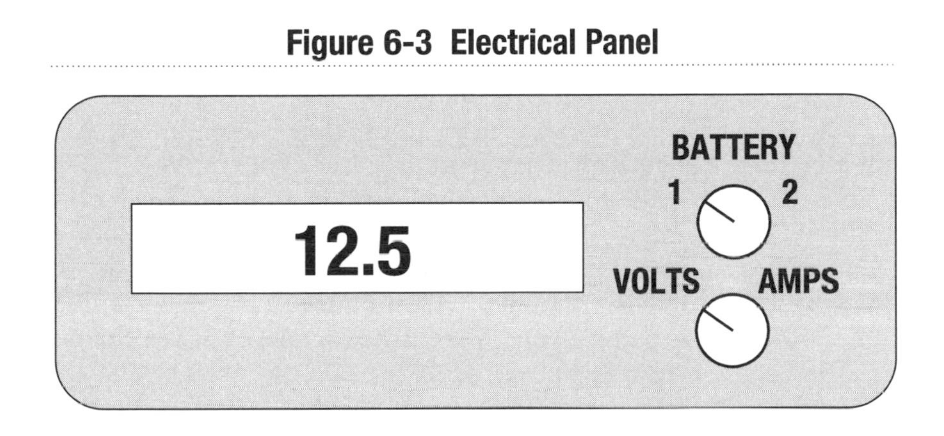

In figure 6-3, by turning the top knob to battery bank #1 and the bottom knob to volts, the display indicates battery #1 voltage. By turning the bottom knob to amps, the display indicates the amperage supplied by battery bank #1 to a load. The same information can be obtained on battery bank #2 by turning the top knob to battery #2. The electrical panel provides an easy and convenient way to monitor your 12 volt electrical system.

Some of the panels are very technically advanced. One, called the amp-hour meter, comes very close to being an "electrical fuel gauge." An amp-hour meter adds amp-hours as the battery is being charged and subtracts amp-hours as the battery is being discharged. For example, an amp-hour meter indicating a minus 50 means that 50 amp-hours have been removed from the battery. As the battery is charged, the reading decreases until it reads 0. This indicates the battery is fully charged. When you turn on electrical loads, the amp-hour meter indicates amp-hours removed from the battery. For example, if you watched a 12 volt 5 amp TV for one hour, the amp-hour meter would display a minus 5, indicating that after one hour you have removed 5 amp-hours from the battery. When you use this meter, it is easy to determine when to charge and when to stop charging your batteries. The amp-hour meter displays information enabling you to manage successfully your battery resources.

WIRING

Upgrading your charging system and battery capacity may require that the electrical wiring be upgraded as well. The following illustrates this point.

A high pressure water pump may not fill a water tank quickly if a small diameter hose is used. A large diameter hose must be installed that will not restrict the water flow to the tank. The larger the hose, the greater the flow will be, up to the maximum capacity of the water pump. Also, the longer the hose the more the resistance the hose has on the water flow.

The wiring for a 3 amp battery charger is quite small because of the small amount of current it carries. Upgrading the charging system to a 40 amp multi-stage battery charger requires a much larger diameter wire. If the same size wire is used with the 40 amp charger, the small diameter wire provides considerable resistance to the higher current. In an electrical circuit, an increase in resistance produces a voltage drop and heat, possibly to the point where a fire starts. An undersized wire may not create a fire, but it causes marginal performance in the electrical system.

All wire inherently produces resistance in an electrical circuit. The smaller the diameter of the wire and the greater the distance between the battery and the electrical load, the greater the resistance in the circuit. In an electrical circuit, a voltage (pressure) drop occurs as the distance from the power source and the load increases. Some electrical devices, such as lights, can function adequately at lower voltage, but electronic equipment and some electrical motors must have voltage within a narrow tolerance. A 3 percent voltage drop is normal in long runs of wire. A 3 percent voltage drop results in a 0.4 volt drop when a charging device is set at 13.8 volts. The battery receives only 13.4 volts, which is not sufficient voltage to charge the battery efficiently. Where possible, install the batteries as close to the battery charger as practical, and use the proper size wire.

The wire gauge standard is called AWG (American Wire Gauge) in the United States, where the higher the number the smaller the diameter of the wire. The proper size wire for a 3 amp charger is only a small #18 gauge wire for lengths of 10 feet, but for a 40 amp charger, #8 gauge wire, which is 3 times the diameter of a #18 gauge wire, is required, and even a larger wire would be better.

Table 6-2 shows the proper size wire for 12 volt circuits. The length of the wire must include both the positive wire and the return ground or negative wire. If the distance is 10 feet from the battery charger to the battery, the length of the wire run is 20 feet. The amount of wire is double the distance because you are installing two wires, one for the positive line and one for the ground or negative line to complete the circuit. If the amperage through the circuit is 40 amps, you would need a #6 gauge wire and not the smaller #8 gauge, which would be used if you were only going to run 10 feet of wire.

Table 6-2 Wire Size for 12 Volts with a 3% Voltage Drop

The distance is measured from the battery to the load and back to the battery

	Distance in feet						
Amps	**10**	**15**	**20**	**25**	**30**	**40**	**50**
5	18	16	14	12	12	10	10
10	14	12	10	10	10	8	6
15	12	10	10	8	8	6	6
25	10	8	6	6	6	4	4
30	10	8	6	6	4	4	2
40	8	6	6	4	4	2	2
50	6	6	4	4	2	2	1
60	6	4	4	2	2	1	2/0
70	6	4	2	2	1	0	3/0
80	6	4	2	2	1	0	3/0
90	4	2	2	1	0	2/0	3/0
100	4	2	2	1	0	2/0	3/0

The failure to use the proper size wire is illustrated by the following example. The size of the battery cable connecting a vehicle starting battery to the starter is very large because the battery produces a tremendous amount of current, and the cable must be large enough not to restrict the current needed by the starter. If the starting battery fails to start the engine, jumper cables are required to jump start the car. If you use cheap, rusty, undersized jumper cables, you often fail to start the car, with the jumper cables becoming very hot. The failure to start the car was not due to the good battery but caused by undersized jumper cables, which only produced heat.

The same thing could happen if a small #18 gauge wire is connected to a 40 amp battery charger. The wire could overheat possibly melting the insulation and perhaps start a fire. The very least that happens is that the battery charger performance is degraded and does not charge the battery adequately.

Inverters, which are discussed in Chapter 2, are great for powering AC loads, but they require a large amount of current. A 1000 watt inverter requires 100 amps to operate at full power. With 100 amps flowing through a small diameter wire, it would become very hot, very quickly. Install an inverter close to the batteries, and use large diameter battery cables connecting the two.

Only multi-stranded copper wire should be used in RV and boat installations, for both AC and DC circuits. Solid core wire found in homes should never be used on RVs and boats. The vibration and minute flexing cause the solid core wire to harden and eventually break. These breaks are extremely difficult to locate.

CONNECTORS

A good electrical connection between the wire and the electrical device is extremely important. Many electrical systems fail not because a device fails but because an electrical connection fails. The vibration of an RV or a boat is often enough to loosen a screw on a terminal, causing the wire to fall out. Even if it does not completely fail, the loose connection may not carry enough current to support the electrical load.

When installing equipment, you want to make connections that provide long-lasting and trouble-free service. Each connection should have good mechanical strength, and very little electrical resistance.

The electrical connection must have good mechanical strength. The wire attached to a terminal must be tight and unable to pull away from the terminal. Twisting a stranded wire around a terminal screw is asking for trouble. Some of the wire strands can work loose from the screw, not only jeopardizing the wire's ability to remain connected but reducing the size of the conductive path. A crimped-on connector or a wire inserted into a terminal block and held with a screw is preferable to twisting the wire around a screw.

Not only the correct sized wire must be used for each application, but also the correct size crimp-on connector and terminal block must be used. A connector or terminal block that is too large for the wire will not grip the wire properly, and the wire may slip out. Strands of the wire can be broken, resulting in poor mechanical strength and electrical conductivity if excessive force is used to crimp the connector or tighten the screw.

Ring type crimped-on connectors are preferable to the fork or spade connectors. If the screw becomes loose because of vibration, common on all RVs and boats, a fork connector can work itself free. A ring type connector will still be making contact, until the screw falls out.

Corrosion also is a problem on RVs and boats, and protecting the connectors from corrosion is important. Wires are insulated preventing corrosion, but the connections are exposed, and if care is not taken, corrosion can start. Corrosion increases the electrical resistance, which not only produces heat and perhaps a fire but also has insulative properties, preventing a clear path for the current. The wire can deteriorate also because moisture from the corroded connection can wick its way under the wire insulation.

To prevent corrosion, insulation tape is often used, but inexpensive tape, especially in the marine environment, often unravels into a sticky goo. Heat-shrink tubing, slid over the connection and heated with a cigarette lighter or hair dryer, melts and its layers fuse, forming a seal that is better than insulation tape.

BATTERY SELECTOR SWITCHES

The battery cables can be connected directly to the vehicle loads and charging devices, or a switch or other device can be installed in the circuit, so you will have some control over how the batteries are charged and discharged.

On boats, it is common to connect a couple of batteries or battery banks to a battery selector switch that allows you to connect all banks electrically, or to isolate them; see figures 6-4 and 6-5. The positive terminals of the batteries lead to the battery selector switch.

The switch has four positions:

1. The off position disconnects both battery banks from any load on the vehicle;

2. The #1 position selects the #1 battery bank connected to position one;

3. The #2 position selects the #2 battery bank connected to position two;

4. The Both or All position selects both battery banks and the banks are connected in parallel.

Figure 6-4 Battery Selector Switch

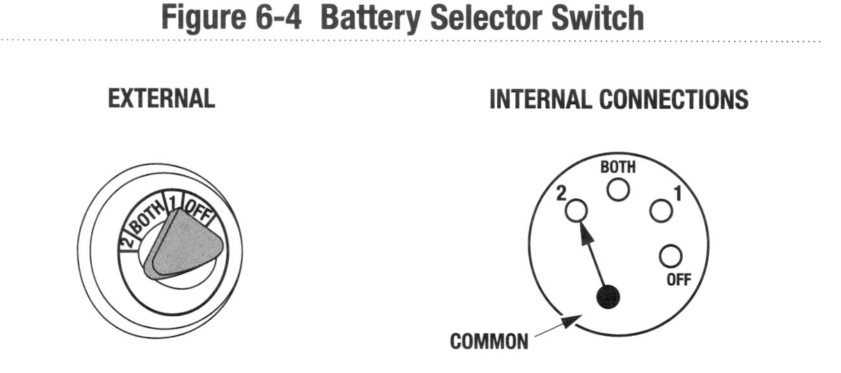

A battery switch allows control over charging or discharging of the batteries. Figure 6-5 shows two sets of 6 volt batteries in series, each with 200 amp-hours.

The switch allows you several options:

First, the battery banks can be isolated from any load on the vehicle by turning the switch to the off position. Before leaving the vehicle, you can turn the batteries off and insure that they are not accidentally discharged because you forgot to turn off some light.

Second, both battery banks can be paralleled by turning the switch to the both position where they act as one large battery of 12 volts 400 amp-hours.

The third option, which makes the battery switch so versatile, is that each battery bank can be isolated from the other.

This enables you to have an easy and a convenient method to switch from one battery bank to the other or connect them in parallel. A spare battery bank is kept in reserve by having only one battery bank selected. If close attention is not paid to the amount of amps being used, you can completely discharge a battery bank. If this happens, just switch to the spare battery bank to continue using the electrical devices on board your vehicle.

Figure 6-5 Battery Banks Connected by a Battery Switch

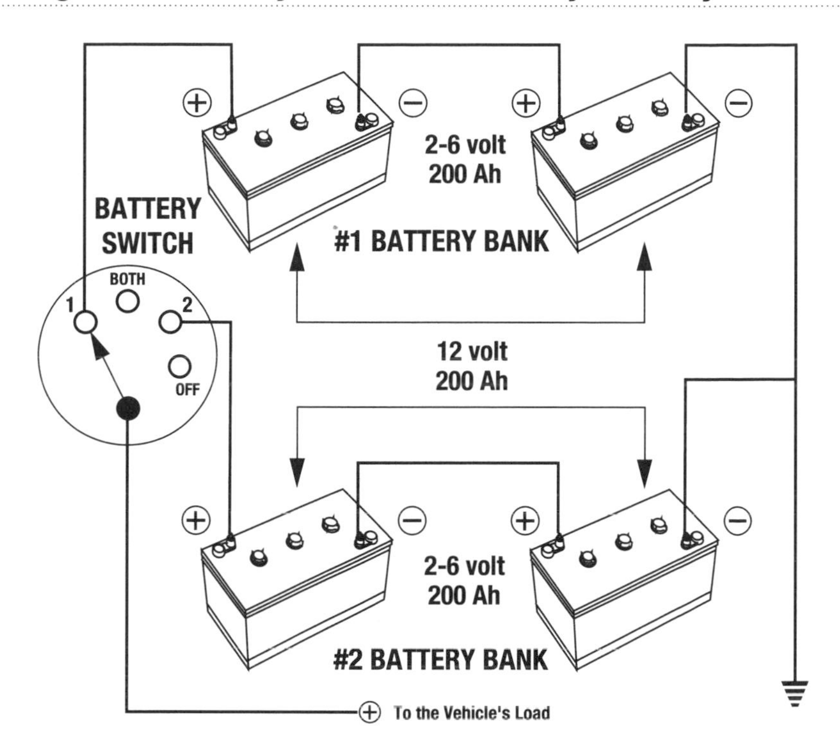

Unfortunately, just like everything else in this discussion of 12 volt electrical systems, the battery switch is not perfect. The battery switch is not automatic. A human must operate it and turning the switch to the wrong position at the wrong moment can sometimes have disastrous results.

If the battery switch is turned off while the alternator is charging the batteries, the alternator and regulator can be damaged. The voltage regulator controls the alternator output by sensing the battery or the system voltage. When the battery switch is turned to the off position, the voltage regulator senses no voltage because the batteries are no longer in the charging circuit and allows the alternator to produce an uncontrollable amount of voltage. A large amount of voltage could destroy the alternator and voltage regulator. Never turn the battery switch to the off position while charging the batteries using an alternator. (This could happen on boats but is unlikely to happen on RVs.) Fortunately, devices are available that attach to the alternator and ground the voltage when the voltage reaches unsafe levels; thus, the alternator is protected.

A lesser problem is if the battery switch is left at the Both positions after recharging. If you did not switch back to just one battery bank, you will drain both battery banks and not have one in reserve. Also, if you fail to turn the switch to a discharged battery during recharging, the discharged battery will not be charged.

When the batteries are paralleled by turning to the Both or All position and the batteries are not equal in state of charge or voltage, the higher voltage battery supplies current to the lower voltage battery until the banks equalize.

ISOLATORS

Battery isolators, figure 6-6, solve the problems associated with the battery selector switch. However, flexibility and energy are lost using an isolator. A battery isolator consists of two diodes and a heat sink. A diode allows electricity to flow in one direction but not in the opposite direction. When current passes through the diodes, heat builds up and is dissipated by the finned aluminum heat sink the diodes are mounted on.

The charging current from the alternator is routed through the isolator instead of a manual battery switch; see figure 6-7. By replacing the battery switch with the isolator, you insure that the switch is never accidentally turned off. Also, both batteries are charged so you don't need to worry about turning the switch to charge the discharged battery. In addition, the batteries are isolated so a fully charged battery can not drain into a discharged battery.

Figure 6-6 Battery Isolator

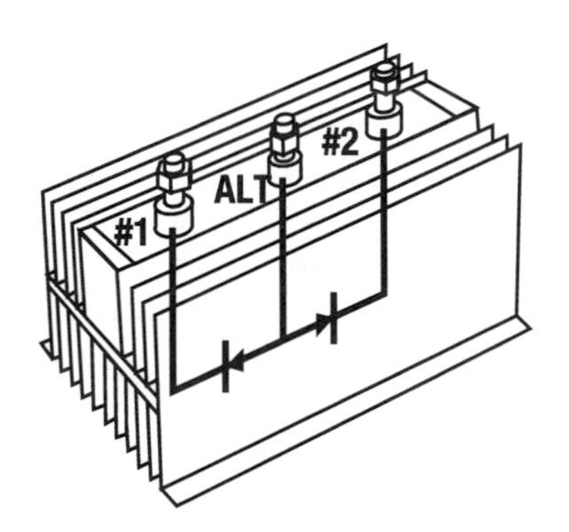

Figure 6-7 Isolator in the Changing Circuit

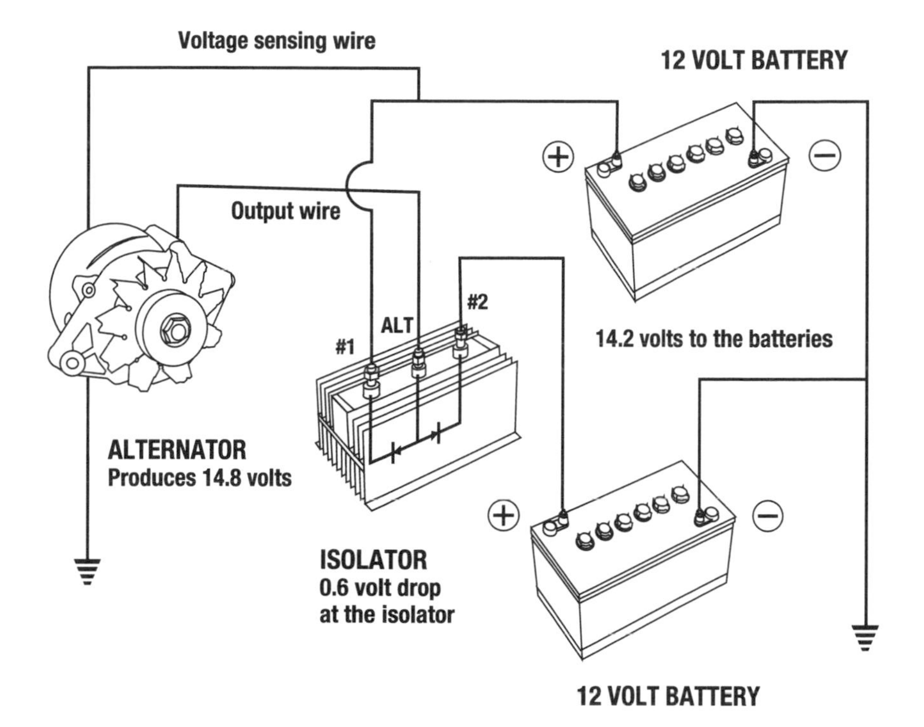

The isolator solves the problems associated with the battery selector switch, but all the advantages of the battery switch are missing. The isolator is inflexible; there are no options.

In addition, energy is lost when using a battery isolator. The diode is not efficient; it causes about a 0.6 volt voltage drop that produces heat. This may not seem like a large drop, but as you learned in Chapter 3, the higher the voltage the more quickly the battery is charged. Most alternators charge at between 13.8 and 14.4 volts and with an isolator installed in the charging circuit the batteries are only being charged at a voltage of 13.2 and 13.8 volts, which is too low to recharge your batteries quickly. The batteries, also, never reach the gassing voltage, so not all the lead sulfate is removed from the plates.

This problem can be solved by rewiring your voltage regulator. Some voltage regulators have an external voltage sensing wire that can be placed on the battery side of the isolator; see figure 6-7. The voltage regulator senses the battery voltage allowing the alternator to produce 14.8 to 15 volts. The batteries will then receive from 14.2 to 14.4 volts after the 0.6 volt voltage drop of the isolator. The batteries are now adequately charged even with the isolator voltage drop. Not all alternators, however, have an external sensing wire. Internally regulated alternators and all one wire alternators may require a wiring change to provide external voltage sensing for the voltage regulator. See an electrical technician to determine if you can rewire your voltage regulator to sense the battery voltage when using an isolator.

SOLENOIDS

On RVs, a solenoid, figure 6-8, is often used to isolate the starting battery from the house batteries. A solenoid is an electrical switch that is activated by an external switch—often the ignition switch. A solenoid consists of a plunger, an electromagnet, and usually two sets of terminals, one small and the other large in size; see figure 6-9. The wires from the ignition switch are connected to the two small terminals, and the battery cables are attached to the large terminals. When the ignition switch is turned on, it energizes the electromagnet, moves the plunger that closes the contactor, and connects the house battery to the alternator and starting battery. When the ignition switch is turned off, de-energizing the electromagnet, a spring in the solenoid pulls open the contactor from the large terminals, and the house batteries are isolated from the starting battery.

The solenoid combines some of the best features of the battery selector switch and the isolator. The solenoid does not have the voltage drop that the isolator has, but it isolates the starting battery from the house battery without having to do so manually, as is the case with a battery

Figure 6-8 Solenoid

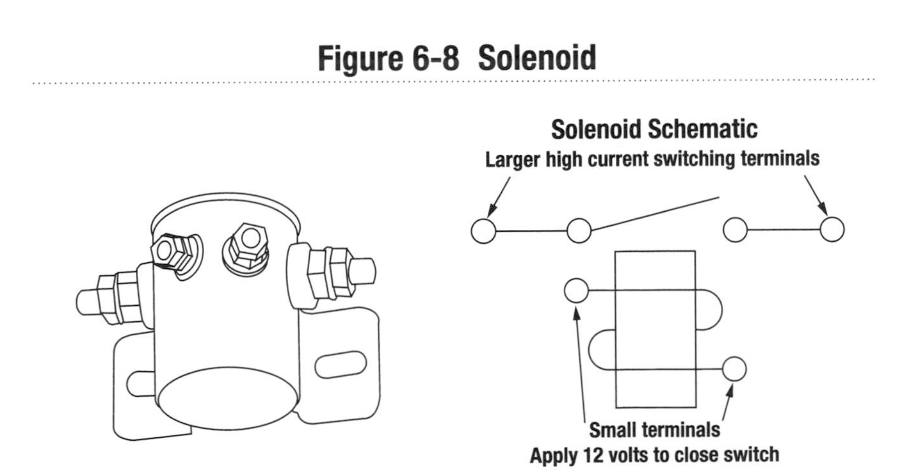

selector switch. When the ignition switch is turned off, the two batteries are isolated.

On some RVs, a second solenoid is installed so that the RVer can isolate the house battery from the loads when he or she leaves the vehicle; see figure 6-10. This has the same effect as turning off the battery selector switch.

Figure 6-9 Cutaway View of a Solenoid

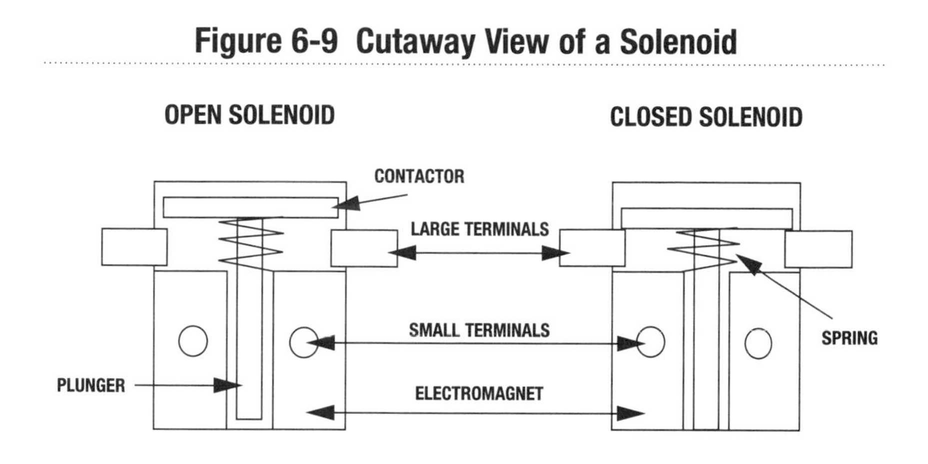

Figure 6-10 Two Solenoids in a Circuit

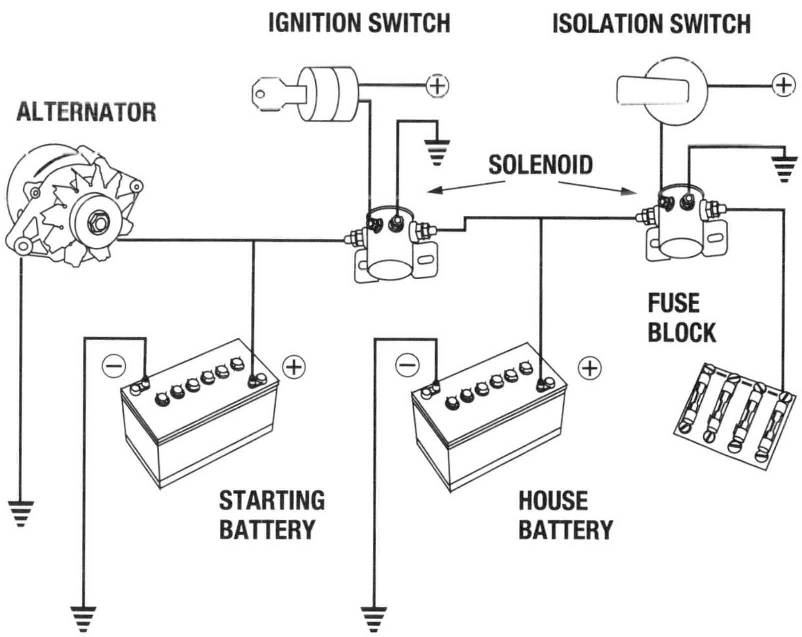

IGNITION SWITCH
ISOLATION SWITCH
ALTERNATOR
SOLENOID
FUSE BLOCK
STARTING BATTERY
HOUSE BATTERY

A solenoid installed to parallel batteries must be a continuous duty solenoid, similar in design and mechanical movement to a starter solenoid, but the starter solenoid is designed for momentary duty and will burn out in continuous duty. Like all other components in an electrical circuit, the solenoid must be designed to handle the maximum current the circuit produces. If undersized or not designed for the application, the component will not only fail but perhaps start a fire.

Chapter 7

Designing and Operating Your 12 Volt System

If your system fails to achieve your needs, redesign your system.

In the previous chapters, you learned how to determine your daily electrical requirement, how the battery works, about different battery types, about various charging systems, and about monitors. In this chapter, this information is brought together and recommendations are made on how to upgrade the electrical systems for various vehicles.

There is not one "best" system that applies to everyone. Each situation is different because of the vehicle type and usage and of financial concerns. But guidelines have been developed in the previous chapters that allow you to determine a workable solution for your 12 volt system. The information about batteries, charging devices, and monitors should allow you to determine what will work best for your situation. The recommendations in this chapter will work for the various vehicles and situations listed, but other designs will work as well.

KEYS TO ELECTRICAL SELF-SUFFICIENCY

The following should be followed:

1) Know your daily electrical requirement.
Spend some time determining your vehicle daily requirement because it is the basis for calculating your battery capacity and charging system. If you do not, it is just luck that your 12 volt electrical system will provide you with reliable service. Use the worksheet at the end of Chapter 2 to determine your daily electrical requirement.

2) Your battery capacity should be three times your daily requirement if you are planning on charging your batteries daily. If you want to go several days between charging, increase your battery capacity.

Determine your battery capacity as follows:
Your daily electrical requirement times 3, times the days between charging, equals your battery capacity.

For example, if you want to go two days between recharging and your daily electrical requirement is 50 amp-hours, you would determine your battery capacity as follows:

50 amp-hours x 3 x 2 days = 300 amp-hours of battery capacity

Table 7-1 lists the battery capacity necessary to support various daily requirements.

Table 7-1 Required Amp-Hour Capacity

	Days Between Charging		
Daily requirement in amp-hours	**One**	**Two**	**Three**
25	75	150	225
50	150	300	450
75	225	450	675
100	300	600	900

If your daily electrical requirement is 75 amp-hours and you want to go two days between recharging, your battery capacity should be 450 amp-hours. A battery capacity of 450 Ah could be achieved by installing two battery banks where each bank consists of two 6 volt 220 amp-hour batteries connected in series. This would give you 440 amp-hours. If you recharged daily, only one of these banks is necessary.

3) Your charging device should produce 25 percent of the battery capacity. In Chapter 3, it was determined that a charging device, alternator or battery charger, can recharge at a rate of 25 percent of the battery capacity when the battery is at 50 percent of charge.

Multi-stage voltage regulation is recommended over a constant voltage or a taper charging device. Standard alternators and battery chargers are regulated by constant voltage regulation. A multi-stage alternator or battery charger produces more amperage during the bulk stage of charging compared to a constant voltage regulated device, thus reducing the time to recharge the batteries.

Determine the output rating of a charging device by multiplying your battery capacity by 0.25.

440 amp-hour battery capacity x 0.25 = 110 amps

Table 7-2 shows what 25 percent of the battery capacity would be for the different battery capacities determined in table 7-1.

Table 7-2 Battery Capacity and Charger Output

Daily require-ment	1 day battery capacity	25% of battery capacity	2 day battery capacity	25% of battery capacity	3 day battery capacity	25% of battery capacity
25	75	19	150	37	225	56
50	150	37	300	75	450	112
75	225	56	450	112	675	168
100	300	75	600	150	900	225

Increasing your battery capacity allows you to install a larger charging device.

For example, by discharging a battery or battery bank to 50 percent of charge, a 100 amp-hour battery accepts 25 amps, a 200 amp-hour battery bank accepts 50 amps, and a 400 amp-hour battery bank accepts 100 amps; four times what the 100 amp-hour battery accepts. If a 25 amp charger is used to recharge a 400 amp-hour battery bank, it will take 4 times longer than if a 100 amp charger is used.

High output multi-stage alternators producing over 100 amps and large inverters with multi-stage 100 amp battery chargers greatly reduce the time to recharge a battery bank compared to a standard low output constant voltage charger.

Solar and Wind Power

Solar and wind power are not taken into account in the above tables. Unless you are located in an area where the sun always shines or the wind always blows, it is difficult to determine how many amp-hours you gain a day from these energy sources. If you do use solar and/or wind power to recharge your batteries, you have in effect reduced your daily requirement and battery capacity requirement by the amount of energy these energy sources supply.

For example, if your daily requirement is 100 amp-hours and your solar panel always produces 15 amp-hours a day, for calculation purposes, your daily requirement is reduced to 85 amp-hours. Your battery capacity can be reduced to 255 amp-hours from 300 amp-hours. The engine driven charger can also be reduced to 56 amps from 75 amps.

A wind generator, in favorable conditions, could produce 50 amp-hours or more per day, reducing your normal daily requirement to 50 amp-hours from 100 amp-hours. You may not want to reduce your battery capacity but to maintain sufficient battery capacity to store the energy the wind generator produces. The wind generator could reduce the need to use an engine driven charger. In this example, by maintaining a 300 amp-hour battery capacity and with the daily requirement reduced to 50 amp-hours, you only need to use your engine driven charger every other day.

CHARGING BATTERIES WHILE OPERATING OTHER LOADS

In all RVs and boats, the charging circuit and the circuit that supplies current to the vehicle loads are not separate; they both have one thing in common—the battery.

In figure 7-1, the charging circuit and the circuit supporting common loads, found in both RVs and boats, are combined. The charging device can be an alternator, a battery charger, a solar panel, or a wind generator. It can also be a combination of several charging devices: an alternator or a battery charger working together with a solar panel to charge a battery. A switch between the charging device and the battery could be a switch built into the charger, a battery selector switch, a solenoid, or an isolator.

Figure 7-1 A Charging Circuit

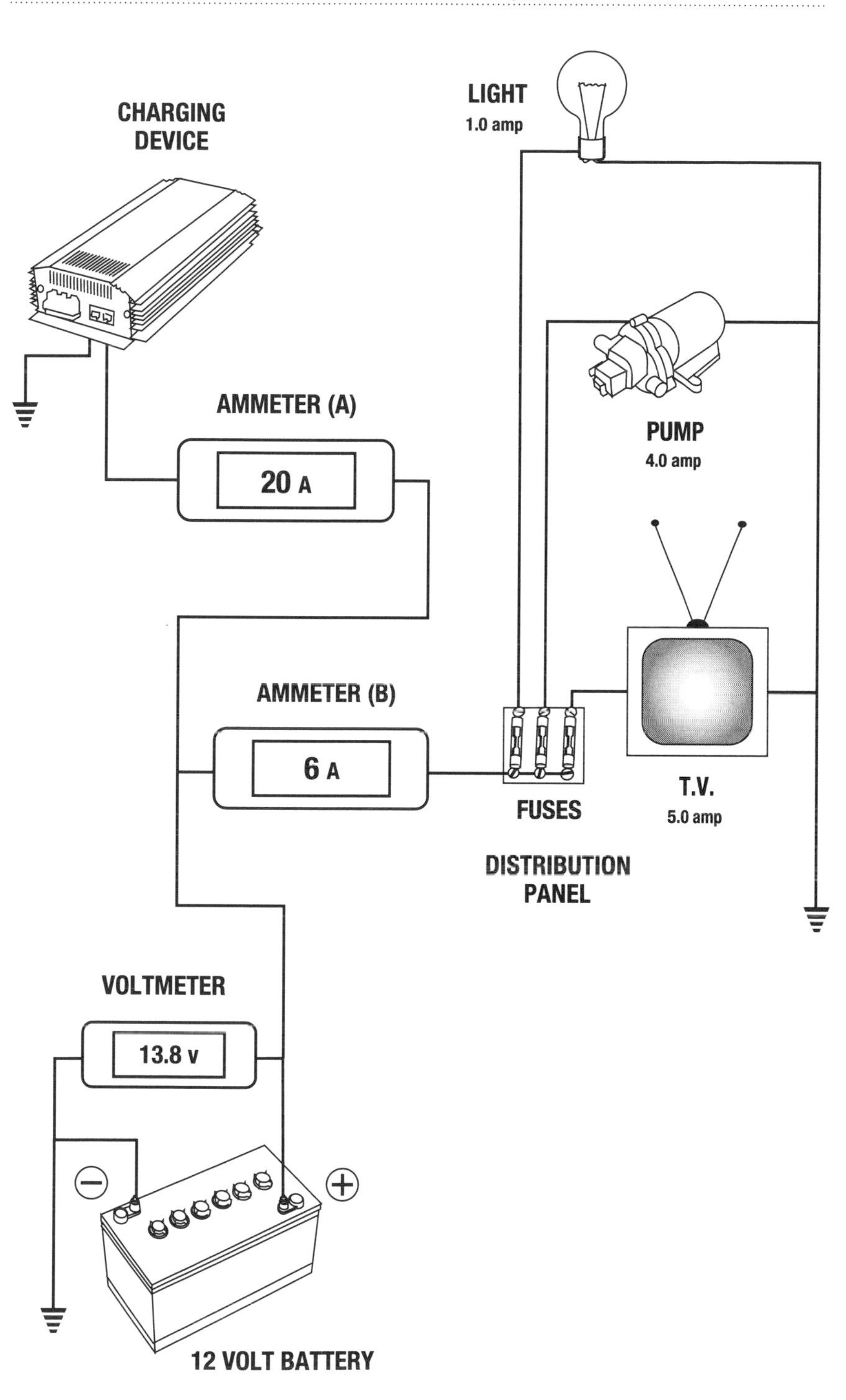

Two ammeters are placed in the combined circuits: one to measure the charging device output (A), and another ammeter to measure the current to the electrical loads (B).

The battery supports the light, pump, and TV when the charging device is turned off.

When the battery needs to be recharged, the charging device is turned on. The charging device not only supplies electrical energy for the battery, but also for the other loads in the circuit.

A charging device is expected to charge the battery and in some cases to do it quickly. What happens to some of that energy if you also have some loads turned on at the same time?

If the battery is a 100 amp-hour battery and it is discharged to 50 percent of capacity it can be recharged by 50 amp-hours before it is fully charged. If your charging device is a multi-stage 20 amp battery charger, the battery can accept the battery charger's full output until it reaches the gassing voltage. The battery charger voltage is much higher than the 12 volts of the discharged battery, so the battery charger is also supplying energy to the other loads in the circuit.

How much of the 20 amps would be used to recharge the battery if you also were watching TV and had the light on? Ammeter A reads 20 amps, and ammeter B reads 6 amps or the power supplied to the 5 amp TV and the 1 amp light. This leaves only 14 amps of the charging current being used to recharge the battery. Since the full output of the charger is not going to the battery, the battery is not being charged quickly.

If the charging device is a 3 amp solar panel, ammeter A would read 3 amps, but ammeter B would indicate 5 amps because you are watching TV. What is supplying the other 2 amps? The battery. The battery is not being charged, but it is also not being discharged by the full 5 amps the battery would have to supply if the solar panel was not helping to power the TV.

In some situations, it does not really matter if a load is on while the battery is being charged. During the absorption stage of the charging process, when the battery is nearing full charge, the battery only accepts a certain amount of amps. The charging device can provide more amperage than the battery can accept, so when other loads are turned on the charging device increases its output, charging the battery and powering the loads. An example is the multi-stage battery charger. If all loads are off and the battery is approaching full charge and accepting only 5 amps, the output of the battery charger is 5 amps. If the TV is turned on, the multi-stage battery charger increases its output to 10 amps and ammeter A indicates the 10 amps the charger is now producing. Ammeter B reads 5 amps, the current to the TV. The load did not subtract from the amperage being supplied to the battery.

The charging device output is not only used to recharge the battery, but it also provides power to the vehicle's other loads.

RECOMMENDATIONS FOR UPGRADING 12 VOLT ELECTRICAL SYSTEMS

The following section contains four recommendations on how to upgrade the electrical equipment on various RVs and boats, so each will have reliable 12 volt power.

Small RV, Trailer, or Pickup Truck with Camper

Situation:

The owner occasionally uses the camper for dry camping. He/she does not want to spend the money to upgrade the vehicle with multi-stage chargers, solar panels, and monitors, but does want to have enough power to enjoy the camper while dry camping.

Standard Inventory:

Alternator: A 55 amp constant voltage alternator charges the batteries while the engine is running.

Battery charger/converter: The vehicle may or may not have an onboard generator powering the 3 amp battery charger included with most RV converters.

Batteries: The camper has one starting battery for the engine, and a 100 amp-hour house battery. A solenoid or an isolator prevents the starting battery from being discharged by the house battery.

The following is an example of a small RV's daily electrical requirement:

Daily requirement:

DC Devices	Amps	Hours Used	Total Amp-Hours
Lights (6)	1.5 each	0.5 each	5
Stereo	2	4	8
DC TV	5	4	20
Water Pump	4	0.25	1
	Total Minimum Daily Req.		34 amp-hours
Heater fan	7	3	21
	Total Maximum Daily Req.		55 amp-hours

Recommendation:

Battery capacity:

The battery capacity should be increased. The 100 amp-hour battery is three times the minimum daily total, but is too small for the maximum daily total. Adding one more 100 amp-hour battery increases the total battery capacity to 200 amp-hours. Plus, by adding another 100 amp-hour battery allows you to have two days of power for a weekend camping trip, when using only the minimum daily requirement.

Caution:

The easiest way to increase capacity is to purchase another 100 amp-hour battery and parallel it with the old one, but the old battery may have an unfavorable effect on the new battery. The new battery discharges into the older battery trying to charge it if the voltages are different. The weakest battery determines the strength of the two batteries, and during charging the old battery consumes a greater proportion of the current available, to the detriment of the new battery.

You will have much better result by installing two new batteries of equal size and model from the same manufacturer. To keep the installation simple, parallel the two 100 amp-hour batteries and then a battery switch is not required. See Chapter 4 on adding battery capacity.

Figure 7-2 is a schematic of an RV with the converter/battery charger and three batteries: one starting battery and two house batteries.

Charging Cycle:

Weekend camping:

A battery capacity of 200 amp-hours is enough power to supply the electrical requirements for two nights. During a two night camping trip, the batteries do not have to be recharged while dry camping if the batteries are healthy and charged before the first night.

Before leaving home, check the battery voltage, specific gravity, and electrolyte level. Insure that the voltage is 12.6 and the specific gravity is 1.260 or higher. If the readings are low, recharge the batteries. Undercharged and/or dying batteries will not provide good service at the campground, especially if you do not have an adequate charging system.

The batteries will be charged by the alternator during the drive to the campsite.

Figure 7-2 Schematic of an RV

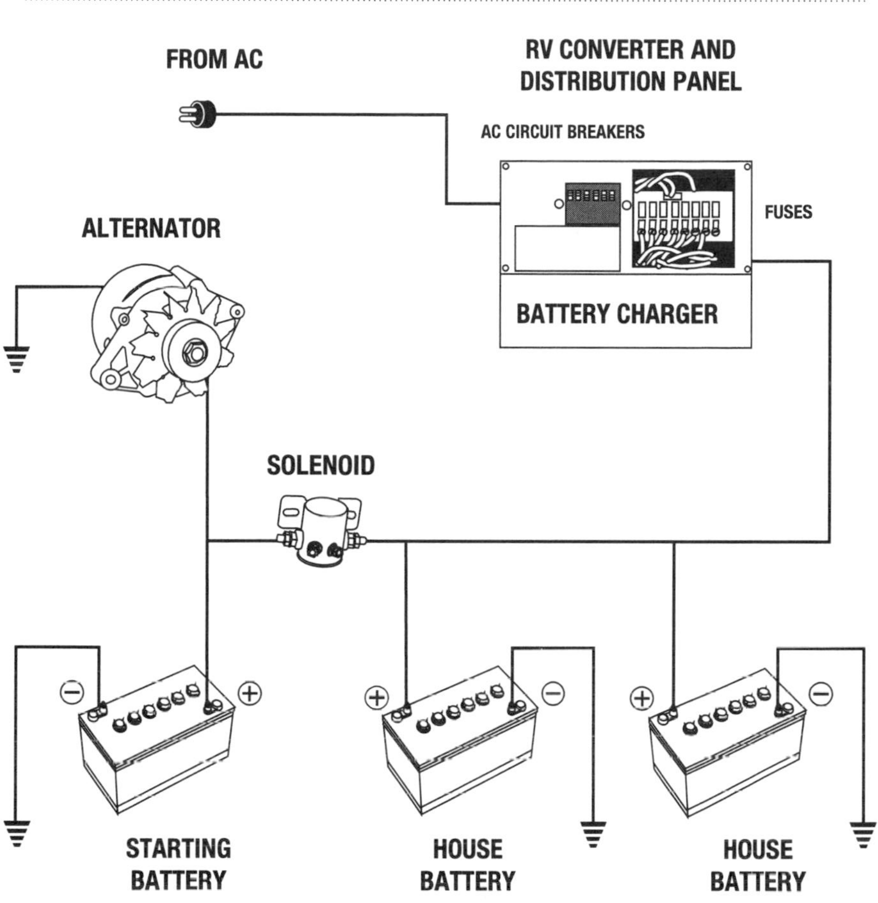

A week's vacation:

During a weeks vacation, the batteries need to be recharged. Without 12 volt monitors, it is difficult knowing when to recharge your batteries to support your daily electrical requirement, so you have to experiment. What do you do?

1. Use a battery charger

If you have an onboard generator and a 3 amp battery charger, which is normal on most RVs, use it to recharge the batteries after the first night. Remember, amp-hours are replaced by amperage over a period of time. It takes at least 12 hours a day using a battery charger with an output of 3 amps to replace a daily requirement of 34 amp-hours.

Plus, the battery must be recharged by an additional 20 percent because of the battery inefficiency.

If an onboard or portable generator is available, a portable automotive battery charger (found at auto parts stores) rated at 10 or more amps may be used to recharge the batteries. These chargers are taper chargers and only produce the 10 amps when the battery is deeply discharged. The current is reduced to 6 to 8 amps as the battery voltage increases, making them only a little better than the converter/battery charger found on most RVs. Some of these chargers have a manual or boost setting where the charger output is unregulated allowing them to produce considerably more amperage. Attaching a deeply discharged battery to a high output unregulated battery charger recharges a battery more quickly than if the automatic setting is used. Many portable battery chargers have a built-in ammeter, so it is easy to determine the amperage supplied to the battery.

Warning:
As the battery reaches full charge, its voltage increases to above 14 volts and battery damage could result if the unregulated battery charger is not turned off. Use a multimeter set at DC volts to monitor the battery voltage when using a non-automatic charger. The voltage initially is 13 volts for a deeply discharged battery climbing to 14 volts. Once the voltage reaches 14 volts, turn off the charger.

2. Use the engine alternator

The RV's alternator produces more amperage to your deeply discharged batteries than the standard battery charger found on most RVs. If your house batteries are discharged to at least 50 percent of charge, the standard alternator charges your battery with about 15-20 amps until the batteries approach full charge. The alternator output then decreases rapidly to a few amps. Using a multimeter set to DC volts, measure the house battery voltage. When the open circuit voltage is about 12.2 volts or the batteries are not supporting the electrical loads, start your engine. The house battery voltage should increase to 13.6-14 volts. You want to run the engine until the alternator output drops to about 7 or 8 amps; at this point it does not produce enough amperage to make running the engine worth while.

Unfortunately, you probably do not have any way to monitor the amperage output of the alternator, so you have to experiment. Run the engine for a few hours. That night, use the lights, TV, and stereo, as usual. If the batteries support the loads okay—the TV is bright with clean images, the lights stay bright—then running the engine for a few hours a day supports your daily requirement. If not, you need to run the engine longer or cut down the time the loads operate—less TV.

A solenoid or isolator is placed in line between the starting battery and the house batteries so the starting battery is not accidentally discharged. The solenoid or isolator is usually located under the hood of the vehicle. If an isolator is in the charging circuit, a voltage drop of 0.6 to 1.0 volt prevents the alternator from charging the battery quickly; in this case, using the alternator to recharge the house battery is not that useful.

3. Use both—the alternator and the generator.
Use the battery charger for a few hours each day, and the engine only when the battery fails to support the loads. If you start with batteries that are in good shape and fully charged, depending on your daily requirement you should have enough energy to last at least a couple of nights. When the batteries become deeply discharged, they will not support the loads unless adequately charged. If you have a low output charging device, charge your batteries each day, for if you don't, you may never catch up.

An RV with an Onboard Generator

Situation:

In this example, an RV is used on weekends to dry camp at the RVer's favorite state park and is also used several weeks each year at a beach that has no facilities. The RVer wants to minimize the time spent at commercial campgrounds, so he/she will purchase and install the electrical equipment necessary to upgrade the 12 volt electrical system. The owner has several 120 volt AC appliances aboard but has not decided whether to install an inverter or to just run the generator each time the AC appliances are used.

Standard Inventory:

Alternator: A 55 amp constant voltage alternator charges the batteries while the engine is running.

Battery charger/converter: The vehicle has an onboard generator powering the standard 3 amp battery charger found on most RVs.

Batteries: The RV has one starting battery for the engine, and a 100 amp-hour house battery. A solenoid or an isolator prevents the starting battery from being discharged by the house battery.

The following is an example of an RV's DC and AC daily electrical requirement:

Daily requirement:

DC Devices	Amps	Hours Used	Total Amp-Hours
Lights (8)	1.5 each	0.5 each	6
Stereo	2	4	8
DC TV	5	4	20
Water Pump	4	0.25	1
	Total Minimum Daily Req.		35 amp-hours
Heater fan	7	3	21
	Total Maximum Daily Req.		56 amp-hours

The AC requirement is as follows:

AC Appliances	Watts	Hours Used	Total Watt-Hours Used
Microwave	900	0.33 (20 min)	300
Blender	300	0.25 (15 min)	75
Coffee Maker	600	0.50 (30 min)	300
		Total AC requirement	675

If an inverter is used to power the AC loads, the following calculations are used to determine the battery capacity needed to support the total load:

Total AC requirement of 675 watt-hours divided by 12 volts = 56 amp-hours.

Since inverters are approximately 20 percent inefficient, the amp-hours must be multiplied by 1.2 to determine the amount of amp-hours necessary to power the AC appliances:

56 amp-hours x 1.2 = 67 amp-hours

Therefore, 67 amp-hours are required to operate the inverter.

The amp-hours required to operate the daily inverter requirement is in addition to the other DC requirements.

The following is the total of both the DC and inverter requirement:

	Amp-Hours
Minimum Daily Requirement	35
Inverter Requirement	67
Minimum Daily Requirement	102
Propane heater fan	21
Maximum Daily Requirement	123

Recommendations:

Battery capacity:

The battery capacity must be upgraded if the RVer is planning to spend weeks at a time dry camping. If the RVer decides not to install an inverter, the maximum daily requirement is 56 amp-hours. Paralleling two 12 volt 100 amp-hour batteries would supply enough power for this situation. It would be better to remove the 12 volt house battery and replace it with two 6 volt 220 amp-hour batteries in series. This would increase the battery capacity and provide true deep cycling capabilities. See Chapter 4 on battery types and how to connect batteries.

If the RVer decides to install an inverter, the maximum daily requirement is about 120 amp-hours. Since the battery capacity should be three times the daily requirement, 360 amp-hours is necessary. Two banks, each consisting of two 6 volt 220 amp-hour batteries in series is preferred. The total battery capacity of 440 amp-hours allows additional equipment to be added to the RV without additional battery capacity.

A battery switch installed between the battery banks increases flexibility on the way the two banks are charged and discharged.

Battery charger:

Install a multi-stage battery charger to replace the 3 amp charger found with most RV converters. If the owner does not plan to use an inverter to power the AC appliances, install a multi-stage battery charger to recharge the batteries quickly and efficiently. Charger models are available with outputs of 20 and 40 amps, so for this situation the 40 amp model is used. The 40 amp model is less than 25 percent of the 220 amp-hour capacity, but its output is sufficient to recharge the batteries quickly. Figure 7-3 is a schematic of an RV with a multi-stage battery charger. Disconnect the 3 amp converter/battery charger when the multi-stage battery charger is installed.

An inverter with a multi-stage battery charger can be installed to power the AC appliances and to charge the batteries. A 1000 watt inverter with a 50 amp multi-stage battery charger would power all the AC appliance in the above example. If you wanted to operate AC appliances requiring additional power, a 2000 watt inverter with a 100 amp multi-stage battery charger could be installed. This inverter would also provide more amperage to charge the batteries.

Figure 7-4 is a schematic of an RV with an inverter. How the inverter is installed in the RV depends on the inverter wattage and what AC circuits it is to power. The inverter's installation manual lists different wiring options, and you will be able to select the proper one for your situation. In figure 7-4, the inverter draws its AC power from a circuit breaker on the AC distribution panel. The inverter only supplies power to one AC outlet on the RV so that a blender or other AC appliances can be operated. The inverter positive DC wire connects to the battery switch. When the inverter is supplying power to the AC outlet, this DC wire supplies power from the batteries. When the inverter is receiving AC power, while plugged into an electrical utility or the generator is running, the multi-stage battery charger, incorporated in the inverter, is charging the batteries through this DC wire. Disconnect the 3 amp converter/battery charger.

A fuse should be installed in the positive cable within 18 inches of the battery post. The fuse protects the battery cables against a short circuit. A 1000 watt inverter requires a 200 amp class T fuse, and a 2000 watt or 2500 watt inverter requires a 300 amp class T fuse.

Solar panel:

A solar panel can be installed to charge the batteries while the RV is stored. When dry camping, the solar panel amperage reduces the time the battery charger needs to be run. Diodes are inline to isolate the batteries from the solar panel after dark. Fuses are also placed inline to protect against a short circuit. The solar panel output is low, but a short circuit carries the full battery current and that can melt the wire and perhaps start a fire.

Monitors:

Sophisticated monitors are available, and many of the advanced inverters have as optional equipment remote panels that display the system voltage, the amperage the batteries supply to the loads, and the battery charger output amperage. Some monitors are close to being a "fuel gauge" for your batteries. Sophisticated panels calculate the amount of amp-hours you have consumed so when you have used 50 percent of your battery capacity you know it is time to recharge. When you recharge, the panels calculate the amp-hours you have restored to the batteries so you will know when to stop charging.

Figure 7-3 RV Schematic with a Multi-stage Battery Charger

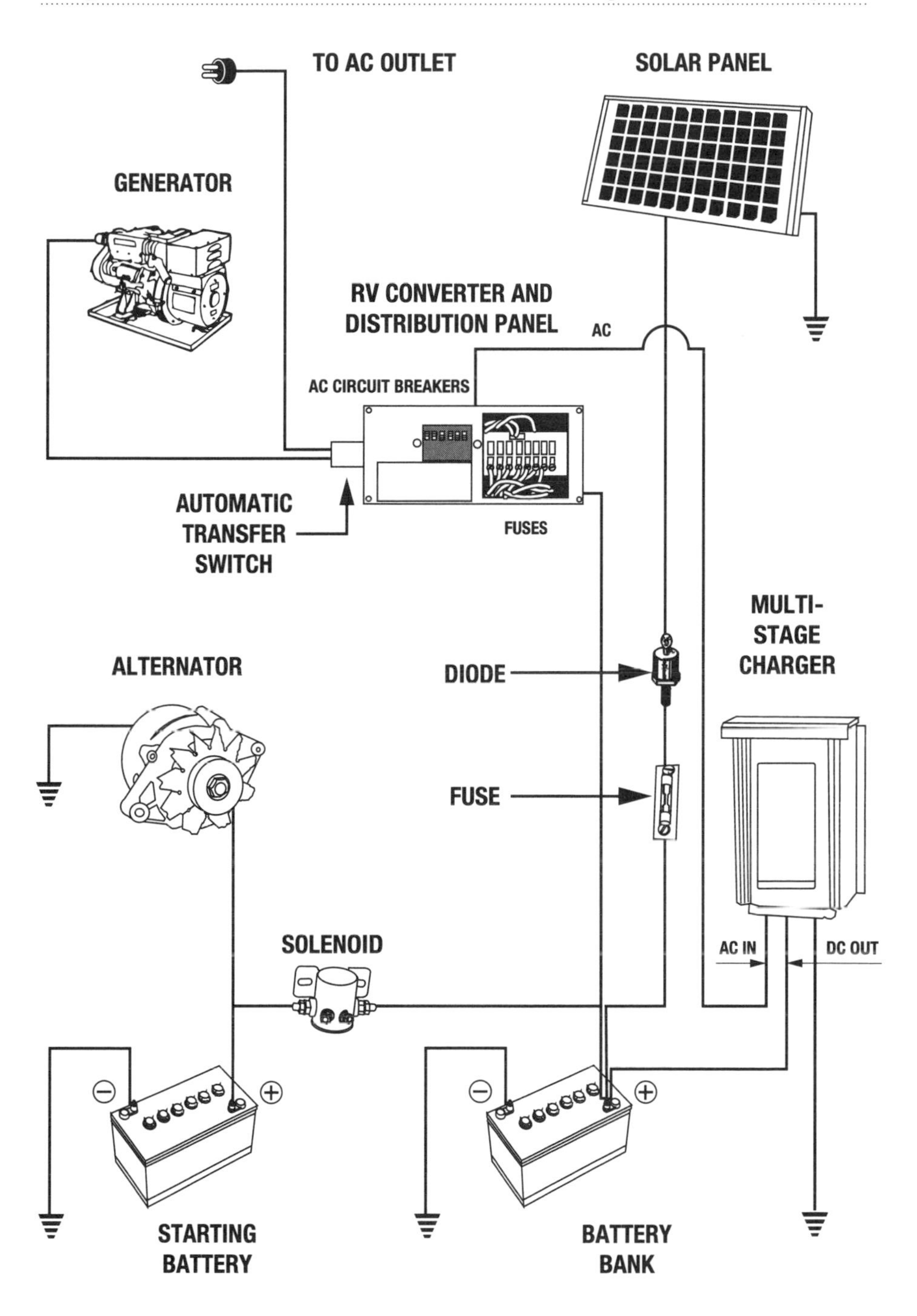

Figure 7-4 RV Schematic with an Inverter

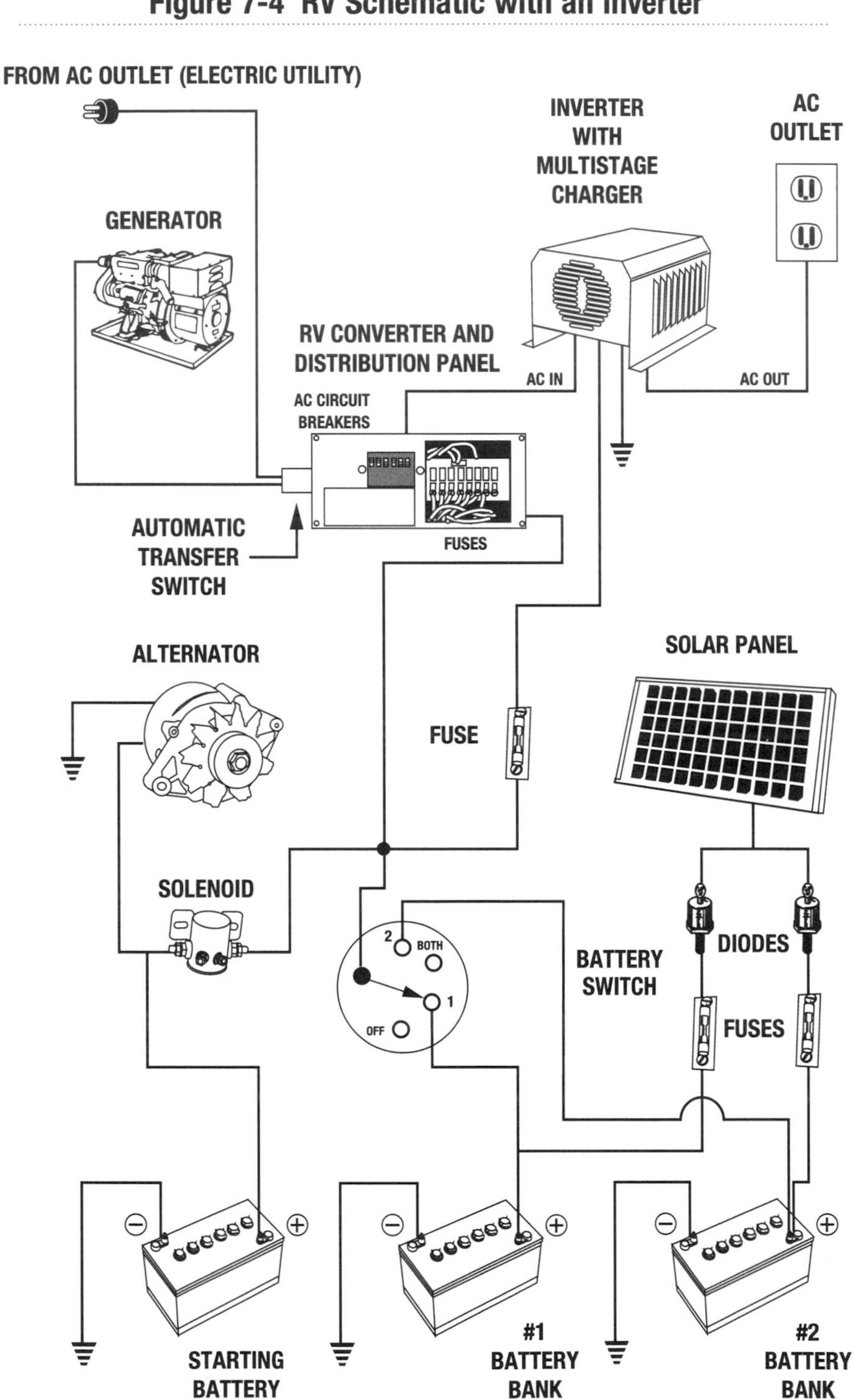

Charging Cycle:

Before leaving home, check the battery voltage, specific gravity, and electrolyte. If the voltage is low, charge the batteries. The batteries are charged by the alternator during the drive to the campsite.

After the first night, the battery capacity is reduced to between 60 and 75 percent of charge depending on the loads used and the time they were in use. If your battery capacity is properly sized, it is not important to have an accurate reading on the amp-hours used. Since you still have power for one more night you can do one of two things, recharge after the first night or wait and recharge after the second night.

If you start charging after the first night, the battery voltage increases quickly to over 14 volts, and the amperage decreases to less than 10 amps. At this point, the battery is charged to about 80 percent and the generator is turned off.

If you wait until the second day to start charging, the battery capacity is reduced to less than 50 percent of charge, depending on the loads in use and the battery capacity. During charging, the voltage slowly increases to 14 volts. The slower the voltage increases the deeper the battery was discharged. Once the voltage approaches 14 volts, the amperage starts to decrease.

Run the battery charger until the amperage drops below 10 amps. The batteries are not fully charged, but they have enough power for another night. It is not worth listening to the generator for the amount of amperage the battery charger is providing to top off the batteries.

Depending on how deeply discharged the batteries and on their health, it should only take a few hours to recharge the batteries to 80 percent of charge. After a few days, you will get a feel for how long you need to run your generator to replenish your batteries' state of charge for another night of camping.

Any time the generator is on insure that the battery charger is charging the batteries. You never know when you will need extra battery capacity for emergency situations.

Periodically, you need to bring the batteries to full charge. This can be accomplished when you drive for many hours, allowing the alternator to bring the batteries to full charge, or when you plug into an electrical utility, allowing the battery charger to top off the batteries. If you purchase a multi-stage battery charger or an inverter that can perform an equalizing or conditioning step, you can keep your batteries healthy by equalizing them once per month.

Battery Selector Switch:

A battery switch is installed between the two battery banks. The following is a guide on how to use the battery switch.

- Off position. The switch in the off position prevents accidental discharging of the batteries.

- Both or All position. The switch should be turned to the both or all position when the engine is running, allowing the alternator to charge both battery banks. Normally the battery switch is in the both position when the battery charger is charging the batteries. Multi-stage battery chargers have charging wires going to each battery bank so that each battery is charged individually. Inverters with multi-stage battery charges have only one positive DC wire, and this wire is connected to the battery switch. When you are plugged into an electrical utility or when your generator is running, the battery switch is typically left in the both or all position so that the inverter's battery charger can charge both batteries.

- 1 or 2 position. After the charging device is turned off, select one or the other battery bank to power the electrical loads. When using the inverter to power your AC appliances, the battery switch is used to select the battery bank used to power the inverter. The other battery bank is in reserve. Discharge one battery bank until it is 50 percent of charge and then switch to the other battery bank. When the second battery bank is at 50 percent of charge, switch to the both or all position and recharge both battery banks.

Small Sailboat

Situation:

In this example, a sailboat owner wants to use the boat to get away to weekend anchorages. During the summer, a two week cruise is planned. During the cruise, the owner plans on anchoring each night so the boat needs to be self-sufficient. The boat has the basic electrical equipment found on most small sailboats, with only a 12 volt refrigerator requiring much power. The boat owner does not plan to operate any AC appliances while away from the marina.

Standard Inventory:

Alternator: A 35 amp constant voltage alternator charges the batteries while the sailboat is motoring.

Battery charger/converter: A constant voltage or taper charger charges the batteries when the boat is dockside. The sailboat does not have a generator.

Batteries: The boat has one starting battery for the engine and a 4D 150 amp-hour house battery. A battery switch is placed between the starting battery and the house battery.

The following is an example of a small sailboat's daily electrical requirement:

Daily Requirement:

DC Device	Amps	Hours Used	Total Amp-Hours
Cabin lights (8)	1.5	0.5	6
Fans (2)	1	3	6
Instruments	1	4	4
Stereo	2	4	8
Water pump	4	0.25	1
VHF radio receive	1	6	6
VHF radio transmit	6	0.5	3
Anchor light	1	10	10
12 volt DC Refrigerator	6	10	60
	Total Minimum Daily Requirement		104
Running lights	3	4	12
	Total Maximum Daily Requirement		116

Recommendations:

Battery Capacity:

The minimum daily requirement is 104 amp-hour and the maximum is 116 amp-hour so the battery capacity needs to be ungraded to three times the daily requirement, 312 to 348 amp-hours for this example. The boat already has one 4D 150 amp-hour battery so an 8D 200 amp-hour battery could be installed next to the existing 4D battery, resulting in an additional battery bank. Insure the old 4D is in good shape before paralleling with the new battery. During charging the old battery takes more of the charging current to the detriment of the new battery. Also, not all 4D and 8D batteries are deep cycle batteries but truck starting batteries. The 12 volt refrigerator requires a lot of energy, and if the batteries are not deep cycle batteries, your electrical system will not provide the service you require.

For true deep cycling, remove the 4D and install two battery banks consisting of a pair of 6 volt 220 amp-hour batteries connected in series. The house batteries now consist of 440 amp-hours which is more than enough power for this example. A starting battery is still used exclusively to start the engine. Battery switches connect the three battery banks.

Battery charger:
Since the boat does not have a generator, upgrading the battery charger to a multi-stage battery charger is not the best option.

Alternator:
Install a 100 amp continuous duty, premium quality alternator with a multi-stage regulator. This allows the boater to replace the daily requirement in only a few hours of engine running time.

Solar panels:
Solar panels are not needed if the boat is berthed in a slip where AC is available. The battery charger keeps the batteries charged. If AC is not available, solar panels can be installed to keep the batteries charged. Solar panels also reduce the time the engine needs to be run while at anchor.

Monitors:
Install a voltmeter to monitor the system voltage, an ammeter to monitor the draw by the electrical loads, and another ammeter to monitor the alternator output. See Chapter Six on monitors.

Charging Cycle:

With a battery capacity of 440 amp-hours, it will take a couple of days before the battery capacity is reduced to 50 percent of charge. Start recharging after the second night at anchor using the multi-stage regulator controlling the alternator.

The alternator output is close to its rated output when charging deeply discharged batteries. The voltage slowly increases to 14 volts depending on how deeply discharged the batteries are. Once the voltage approaches 14 volts, the amperage starts to decrease.

Run the engine until the amperage drops below 10 amps. As the batteries approach full charge, they only accept a few amps. Thus, it takes many more hours to top off the batteries; so turn off the engine. The batteries are not fully charged, but they have enough power to support your requirements for another night.

It should only take a few hours to recharge the batteries to 80 percent of capacity. After a few days you will know how long you need to run your engine so that you will have enough power for another night.

Figure 7-5 is a schematic of a boat with an alternator charging three battery banks.

Figure 7-5 Small Sailboat Schematic

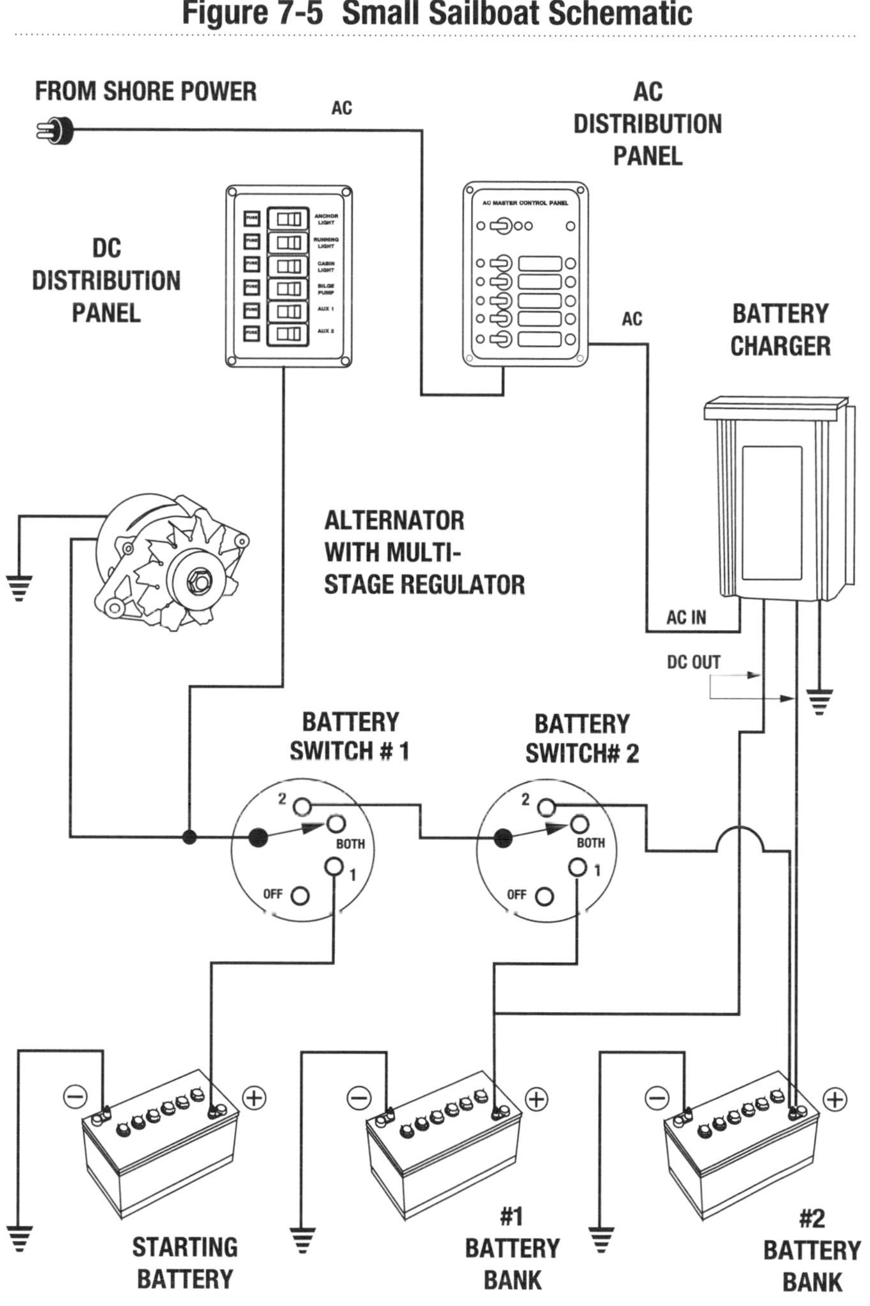

Periodically, you need to bring the batteries to full charge. This can be accomplished by plugging into an electrical utility at a marina allowing the battery charger to top off the batteries, or by motoring, allowing the alternator to charge the batteries.

Battery Selector Switch:
Two battery switches are installed connecting the starting battery and the two house batteries. The following is a guide on how to use the battery switches.

- Off position. Both battery switches can be turned off preventing accidental discharging of the batteries. The wiring to the automatic bilge pumps must be installed before the battery switches so that the bilge pumps operate when the battery switches are in the off position.

- Both battery switches in the Both or All position. When the engine is running, both switches should be in the both or all position so that all three batteries are being charged.

- Starting the engine. When starting the engine, turn the first battery switch to the 1 position to select only the starting battery. After the engine is started, turn the switches to Both or All so that all of the batteries are charged. If the starting battery voltage is very low and unable to start the engine, select one of the house batteries to start the engine. Insure the switch to the discharged starting battery is off, and the starting battery is not in the starting circuit. If one house battery fails to start the engine, select the other house battery. If each battery fails to start the engine, you would parallel all three batteries by placing both battery switches in the both or all position. Two or three deeply discharged batteries that individually fail to start the engine can have enough power if paralleled. The reason you do not parallel the starting battery initially is that the good battery has to provide power to both the starter and to the discharged battery, trying to recharge it. It may not be able to do both.

- Select only one of the house batteries. After the engine is turned off, turn the first battery switch to position 2, removing the starting battery from the circuit. This procedure insures that the starting battery is not discharged. Select one or the other house battery banks to power the electrical loads by turning the second switch to position 1 or 2. The non-selected battery is now in reserve. Discharge the first battery bank until it is at 50 percent of capacity, and then switch to the other battery bank. When it is at 50 percent of charge, recharge both banks.

Large Sailboat

Situation:

In this example, a large cruising sailboat is planning to spend several years cruising, and most of its time will be spent at anchor and not in marinas. The boat has an extensive electronic inventory, and the owner wants to have several ways to recharge the batteries.

Standard Inventory:

Alternator: A 55 amp constant voltage alternator charges the batteries while the sailboat is motoring.

Battery charger/converter: A constant voltage or taper charger charges the batteries when the boat is dockside. The sailboat has a diesel generator.

Batteries: The sail boat has one starting battery for the engine, and a 8D 200 amp-hour house battery. A battery switch is placed between the starting battery and house battery.

Daily Requirement:

The sailboat is a well-equipped boat with several AC appliances and a refrigerator that is cooled by an AC compressor and an engine driven compressor. Either compressor is run for several hours twice a day to freeze the cold plate. The inefficiency of the inverter makes it impractical to use the inverter to power an AC refrigerator, so the refrigerator is cooled when the generator is operating or when the main engine is running.

The following is an example of a large sailboat's DC and AC daily electrical requirement:

DC Device	Amps	Hours Used	Total Amp-Hours
Cabin lights (10)	1.5	0.5	8
Fans (2)	1	3	6
Instruments	1	5	5
Stereo	2	4	8
Water pump	4	0.25	1
VHF radio receive	1	6	6
VHF radio transmit	6	0.5	3
Anchor light	1	10	10
SSB radio Receive	1.5	4	6
SSB radio Transmit	20	0.5	10
Desalinator	4	5	20
	Total Minimum Daily Requirement		83
Radar	4	2	8
Running lights	3	4	12
	Total Maximum Daily Requirement		103

The AC requirement for the boat is:

AC Appliances	Watts	Hours Used	Total Watt-Hours Used
Microwave	900	0.33 (20 min)	300
Blender	300	0.25 (15 min)	75
Coffee Maker	600	0.50 (30 min)	300
		Total AC requirement	675

If an inverter is used to power the AC loads, the following calculations are used to determine the battery capacity needed to support these loads:

Total AC requirement of 675 watt-hours divided by 12 volts equals 56 amp-hours.

Since inverters are approximately 20 percent inefficient the amp-hours must be multiplied by 1.2 to determine the amp-hours from the batteries necessary to power the AC appliances:

56 amp-hours x 1.2 = 67 amp-hours

Therefore, 67 amp-hours are required to operate the inverter.

The amp-hours required to operate the daily inverter requirement is in addition to the other DC requirements.

The following is the total of both the DC and inverter requirements:

	Amp-hours
Minimum Daily Requirement	83
Inverter Requirement	67
Minimum Daily Requirement	150
Running Lights, and Radar	20
Maximum Daily Requirement	170

Recommendations:

Battery Capacity:

Three times the maximum daily requirement is about 500 amp-hours. Two banks of 6 volt 250 amp-hour batteries (a total battery capacity of 500 amp-hours) is enough capacity to power all the electrical devices onboard the boat.

Battery Charger:

Since the boat has several AC appliances, install a 1000 or 2000 watt inverter. Some inverters this size include 50 or 100 amp multi-stage battery chargers. Several installation options are available depending on how many AC appliances you want the inverter to power. The inverter installation manual lists several options, allowing you to select the best one for your situation.

Figure 7-6 shows the inverter receiving AC power from one branch of the AC distribution panel. The inverter is supplying power to only one AC outlet so various AC appliances can be plugged in and operated. When the generator is running, the inverter's multi-stage battery charger is recharging the batteries via the positive DC cable connected to the first battery switch. The two battery switches should be in the All or Both positions to recharge all batteries. With the generator off and the inverter supplying AC voltage to the AC outlet, the battery switches are positioned so that only one battery bank is supplying power. The other battery bank is kept in reserve. Of course, the starting battery is always kept isolated from the other batteries whenever the engine is off.

A fuse should be installed in the positive cable, within 18 inches of the battery post. The fuse protects the battery cables against a short circuit. A 1000 watt inverter requires a 200 amp class T fuse, and a 2000 watt or 2500 watt inverter requires a 300 amp class T fuse.

Alternator:

Install a 100 or 130 amp, premium quality alternator with a multi-stage regulator. By having both a multi-stage battery charger and multi-stage regulator controlling the alternator, the boater has the option of recharging the batteries by running the engine or the generator. This situation provides the boater with two powerful options.

Solar panels:
Solar panels can be installed to charge the batteries during the absorption stage. Two parallel diodes are inline to prevent current from flowing from the batteries to the solar panel after dark. Fuses are also placed inline to protect the circuit from a short.

Wind Generator:
A wind generator could be installed if the boat is going to spend time in the tradewinds. The output of a quality wind generator greatly reduces the energy needed from the batteries, and perhaps reduces the frequency the engine driven chargers must be run. A diode is placed inline so that the battery does not discharge to the wind generator when the wind stops.

Monitors:
Sophisticated monitors are available that display the systems voltage, the amperage the batteries supply to the loads, and the output amperage of the various chargers. Advanced panels calculate the amount of amp-hours you have consumed so when you have used 50 percent of your battery capacity you know it is time to recharge. When you recharge, the panels add the amp-hours you have restored to the battery so you know when to stop charging your batteries.

Figure 7-6 is a schematic of a boat with an inverter and an alternator charging three battery banks.

Charging Cycle:

Since the boat has an AC/engine driven refrigerator system, either the generator or the main engine is run each day to freeze the cold plate. The batteries are charged by the multi-stage battery charger when the generator is run, and by the alternator with a multi-stage voltage regulator when the engine is operating. In this situation, 12 volt battery power should not be a problem, for each time the refrigerator is cooled the batteries are charged.

If the refrigerator is 12 volts, like the example of the small sailboat, additional battery capacity is needed to support the 60 amp-hours the 12 volt refrigerator requires. The minimum daily requirement would increase to about 200 amp-hours per day and the maximum daily requirement to 220 amp-hours per day. Install two banks of large capacity 6 volt batteries with amp-hour capacity of 300 Ah or more. If you

Figure 7-6 Large Sailboat Schematic

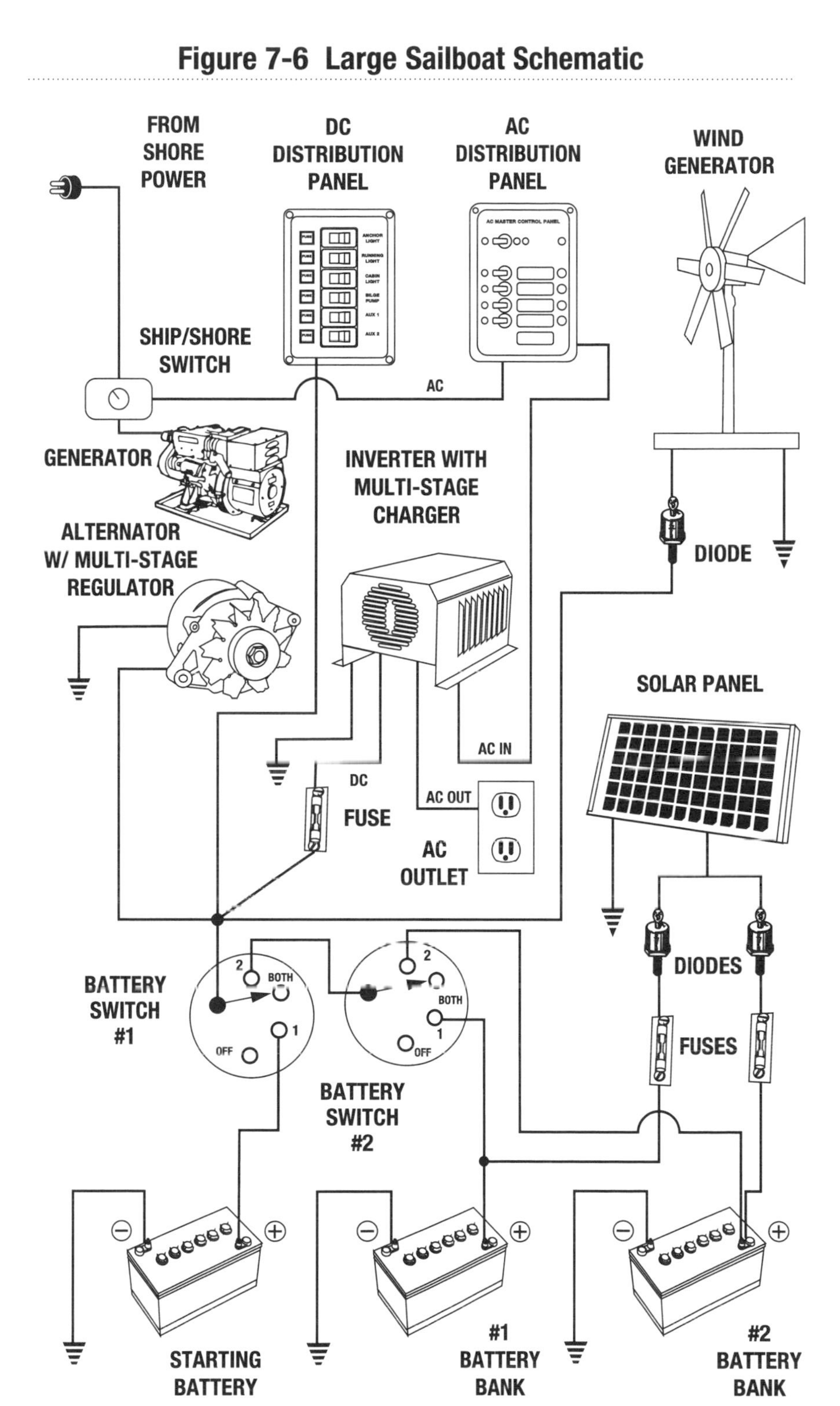

need greater capacity, 2 volt single cells are available from 320 amp-hours to 1000 amp-hours. Use 6 of these 2 volt cells in series to make a very large capacity 12 volt battery bank.

Solar panels and/or a wind generator could reduce the time the engine driven chargers need to be run. The monitors would indicate when the batteries reach 50 percent of charge, at which time the boater would start charging using an engine driven charger. Depending on how much amperage the solar panels or wind generator produce, the boater probably only needs to run the engine driven charger every few days.

With so many options, it would not take the boater long to determine which combination of charging systems he/she needs to operate to keep the batteries charged and supplying all the power the boater requires.

Chapter 8

Understanding Electrical Circuits

If you understand electrical circuits, your troubleshooting will be successful.

It's frustrating having something that does not work properly, and this is especially true of electrical circuits. Many people do not understand electrical circuits and rely on skilled electrical technicians to solve their problems. What is even more frustrating is after waiting for the technician you find that the only problem is a blown fuse or a loose connection. With an understanding of electricity and electrical circuits, and with a multimeter or test light, you can troubleshoot your electrical problems and correct them or determine that the problem is best left for the technician.

A SHORT COURSE IN 12 VOLT ELECTRICITY

Before you can troubleshoot a failed electrical circuit, you must learn about electricity and electrical circuits, and how a good circuit functions. Since you can not see the electricity in a 12 volt system, special tools help you to understand and troubleshoot electrical circuits.

The following is a short course explaining electricity, circuits, and multimeters. The best way to understand an electrical circuit is to practice using a multimeter on known good circuits. This practice allows you to become skilled at finding problems in failed circuits.

ELECTRICITY

Electricity has three components: voltage, current, and resistance. In an electrical circuit, a force pushes the electrical current against an opposition or resistance consisting of the load and material of the circuit. This force or pressure is called voltage. The electrical current the voltage pushes is a flow of free electrons, and the force must be strong enough to overcome the resistance of the load and material of the circuit.

Electrical Current

Electrical current is the flow of free electrons past a point of an electrical circuit. The ampere is the unit of measure of electron flow rate or current (symbol I) through a circuit.

Voltage

Voltage (V), or difference in potential, is the force that propels the free electrons around the circuit.

Difference in potential can be described by the analogy of two water tanks connected at the bottom by a pipe and a valve; see figure 8-1. One tank is filled with water; the other tank is empty. With the valve closed, there is no flow, but the potential for water flow, or a force, exists at the valve. The difference in the water pressure between the two tanks is what causes this force, or difference in potential, to exist at the valve.

Figure 8-1 Potential Difference Between Two Water Tanks

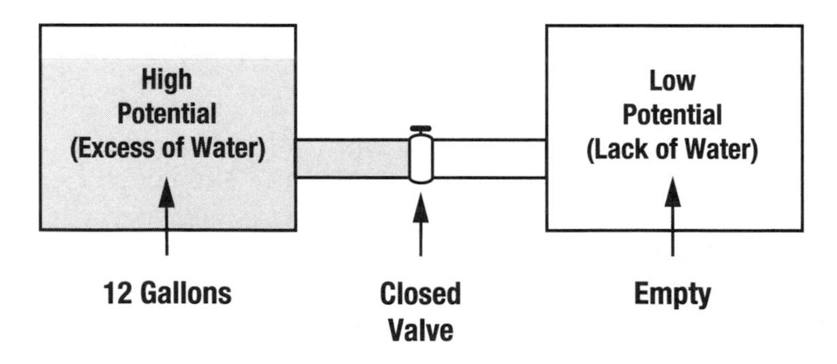

Difference in potential also occurs in a charged lead acid battery. During the electrochemical process (described in Chapter 3), a surplus of electrons at the negative plates results in a higher potential energy level than that at the positive plates. This is similar to the two water tanks, where one tank has more water pressure than the other. At the water tanks, when the valve is opened, the water force is released. In a battery, figure 8-2, when a switch is closed, the excess of electrons flows to an area with a lack of electrons, releasing the electron force to do useful work at the load.

At the water tanks, the force of the water flow decreases as the water levels in the two tanks equalize. In the battery, as the sulfuric acid in the electrolyte is consumed and the plates are covered with lead sulfate, the potential difference or voltage between the negative and positive plates is reduced. The battery becomes discharged and is unable to provide a flow of electrons though the circuit. The battery is recharged by a charging device that reverses the flow of electrons, restoring the potential difference between the negative and positive plates.

Figure 8-2 Potential Difference in a Battery

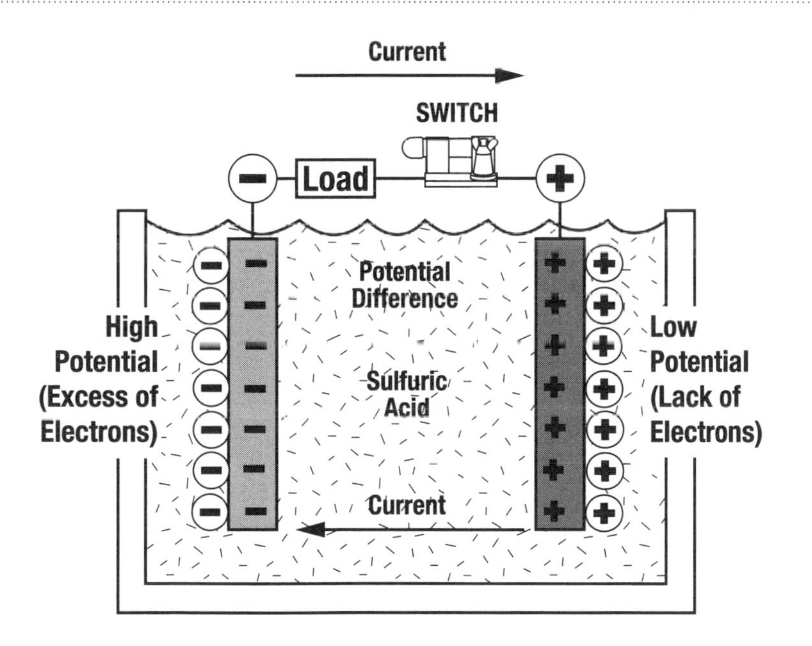

Resistance

The opposition to the flow of electrical current is called resistance. The unit of measurement for resistance is the ohm, and its symbol is R. All materials in a circuit provide resistance to the current. Even copper conductors have some resistance. Insulators are materials with extremely high resistance that prevents the flow of electricity through them.

The analogy of the two tanks connected at the bottom by a pipe and valve can help to explain resistance. A closed valve is similar to an insulator in that it prevents the flow of water. When the valve is opened, the water flow is only restricted by the dimensions of the pipe, which resists the flow of water. If the pipe diameter is small, the water flow between the two tanks is slow. Increase the pipe diameter and the rate of flow increases. The type of walls inside the pipe, whether smooth or rough, also hinders the water flow.

The wiring in an electrical circuit is similar to the pipes in the above analogy. The size, length, and type of wiring add resistance to the circuit. If the wire is too small for the amount of current, the wire adds unwanted resistance to the circuit and becomes hot. Long runs of wiring also introduce unwanted resistance, reducing the amount of voltage supplied to the loads.

Resistance is not the same for all elements. Copper, silver, and aluminum wiring have different resistance. Temperature also affects the resistance of electrical conductors; some metal's resistance increases with temperature.

The electrical load also has resistance. In the analogy of the two water tanks, the water flow is harnessed to do work by placing, after the valve, a paddle wheel that is attached to a machine by a pulley and belt; see figure 8-3. When the valve is opened, the water flow between the two tanks turns the paddle wheel. If the machine is easy to turn, the paddle wheel turns quickly, allowing a large amount of water to pass. If the machine is difficult to turn, the paddle wheel turns slowly and a small amount of water flows.

The load in an electrical circuit acts similarly to the paddle wheel. The load is placed between two areas of potential difference (the positive and negative terminals of the battery in figure 8-2) just as the paddle wheel is placed between the full tank and the empty tank. Once the switch completes the electrical circuit or the water valve is opened, the resistance of the electrical load or the paddle wheel and the dimensions of the wires or pipes determine the amount of current in the circuit. If the electrical load or the paddle wheel turns freely, it provides only a small amount of resistance to the current, so the amount of current is high. If the load has a large amount of resistance, the amount of current is small.

Figure 8-3 A Paddle Wheel Turning a Machine

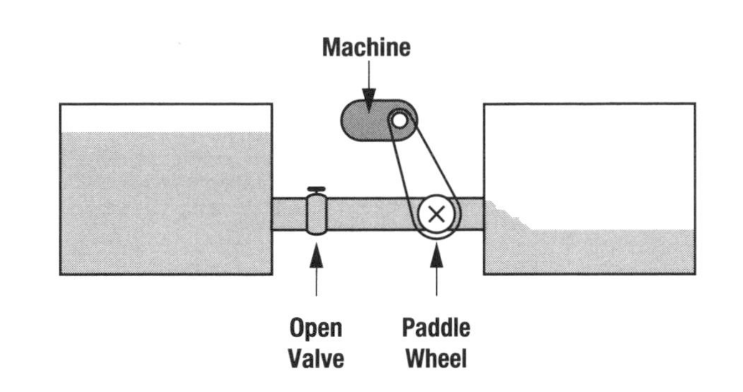

If the voltage or potential difference is constant, the circuit resistance determines the current.

Ohm's Law

A mathematical relationship exists between the amount of voltage, current, and resistance in an electrical circuit. This relationship was first established by Georg Simon Ohm, a German physicist, in 1827 and is know as Ohm's Law.

Ohm's Law is simply:

V (volts) = I (amperes) x R (ohms)

It takes one volt of applied pressure to move one ampere of electrical current against one ohm of resistance; see figure 8-4.

Figure 8-4 Ohm's Law

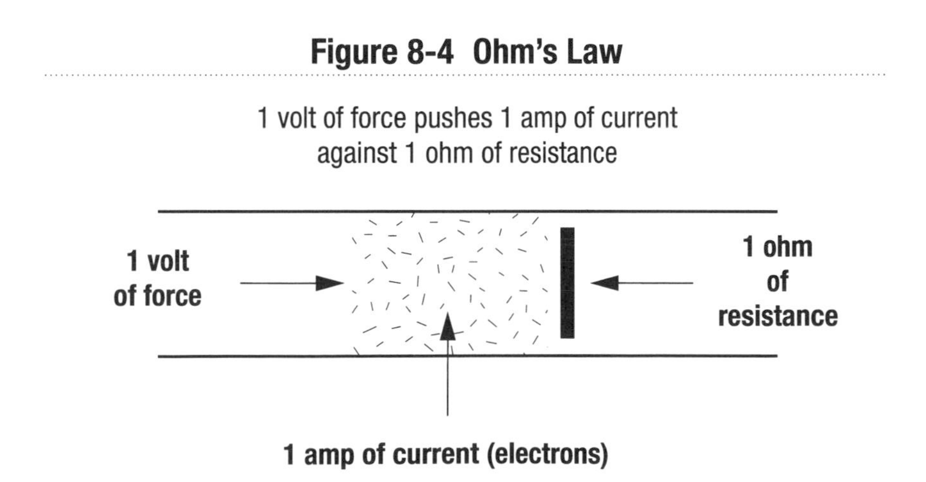

Ohm's law can be written in several different forms:

$$I = \frac{V}{R} \qquad R = \frac{V}{I} \qquad V = I \times R$$

If two of the values are known, the other can be found by using Ohm's law.

These three simple formulas explain how the three components of electricity react in a circuit, and determine how an electrical circuit will perform under different circumstances.

Power is also determined by two of these values. Power, measured in watts, is the rate at which work is being done, or whenever a force causes motion. Work can be mechanical as when a force is used to lift or move a weight, or electrical as when electrons are used to turn a motor. In electricity, power is the force or voltage of a circuit multiplied by the current through the circuit. Or simply:

$$P = V \times I$$

MULTIMETERS

You can only see electricity as lightning or as a spark, or feel electricity when shocked, so a multimeter or a test light is necessary in troubleshooting electrical circuits. Multimeters are instruments that measure voltage, resistance, continuity, and current. A test light checks for voltage in a circuit.

There are two basic types of multimeters: analog and digital. Analog multimeters, the type with a needle, are not as accurate as digital multimeters and are more difficult to understand. A digital meter is accurate to about 1.5 percent and is better protected against misuse than an analog meter.

Small digital multimeters are available that make measuring voltage, resistance, and current in a circuit extremely easy. At a cost of about $25 and up, these multimeters can help you solve some of your electrical problems.

The digital multimeter has a digital display and controls for measuring voltage, resistance, and amperage. There are two probes: one red and one black. The probes on the simplest meters can not be moved. On more expensive meters, the probe leads must be inserted into jacks, and different kinds of leads can be used.

Read the instruction manual before using the multimeter, because the controls, capabilities, and parameters for each meter are different:

some meters can not measure current, others are not intended for AC voltages, and on some meters, the test leads must be repositioned for various readings: read your instruction manual.

TEST LIGHT

A test light is an inexpensive tool for testing electrical circuits. You can make a test light by soldering two wires to a light socket or directly to a light bulb. Solder one wire to the tip at the bottom and the other wire to the base. The wires should be a couple of feet long with insulated alligator clips soldered at the ends. Some test lights have a battery and check for continuity; other test lights measure 120 volts AC, but not 12 volts. You can purchase these test lights at automotive or electrical stores.

PRACTICE USING A MULTIMETER OR A TEST LIGHT

Like any tool, you must learn how to use the multimeter or test light—practice with it. It is not the time to learn how to use the multimeter when a circuit has failed, but when you have time to practice with it and understand what readings you should expect from a good circuit.

Use a multimeter and/or test light and perform the following exercises. You will gain invaluable experience that makes troubleshooting easier.

Measure DC Voltage

Use a multimeter to measure the voltage of a 12 volt DC battery that is not in use:

1. Set the control on DC volts and if the meter does not have auto ranging, set it to read 20 volts; see figure 8-5A.

2. Insert the black lead into the negative or common jack and the red lead into the positive jack on the multimeter.

3. Touch the red probe to the battery positive terminal and the black probe to the negative terminal. The digital readout will indicate

the battery voltage, 12.39 volts in this example. This is the open circuit voltage of the battery.

4. The circuit polarity is determined by the plus or minus reading when measuring DC voltage. If you are using a digital meter, reverse the probes. Touch the black probe to the battery positive terminal and the red to the negative terminal. Reversing the probes does not harm a digital multimeter. The readout now indicates a negative (-) 12.39. On an analog meter, the needle drops quickly below zero if the probes are reversed, and the meter could be damaged.

When measuring the voltage at the load or other circuit components, the polarity of the readout indicates which connection to the component is the positive side and which is the negative side of the circuit. The positive side of the connection leads back to the battery positive terminal, and the negative side leads to the battery negative terminal or to ground. Knowing the polarity is important when installing certain loads: electronic equipment can be damaged, and motors run backward if the polarity is reversed.

Using a Test Light

A test light can determine if a battery has voltage. Attach one clip of the test light to one battery terminal and then clip the other one to the other terminal. If the battery has voltage, the test light will light. The light's brightness gives a clue as to the battery's state of charge. A bright light indicates that the battery has a high state of charge, and a dim light indicates the battery needs to be recharged. The test light is not as accurate as the multimeter, but it can be helpful and is easier to use.

Measure AC Voltage

Use a multimeter to measure the voltage of a 120 volt AC Outlet:

1. Set the meter controls to AC volts, and if the meter does not have auto ranging, set it to read 200 volts; see figure 8-5B.

2. The test lead position is usually the same as when measuring DC voltage, but check the multimeter instructions for correct placement of the leads.

3. Carefully insert the probes into the slots of an AC outlet. No, you will not get shocked as long as you hold onto the plastic probes and not the metal tips. In this example, the voltage is 115.4 volts.

4. Reverse the probes. The voltage is still 115.4 volts. There is not a plus or minus for AC voltage, so it does not matter which probe goes into which slot.

Using a Test Light

A test light rated for 120 volts AC can be used to determined if AC voltage is present. Caution: A 12 volt test light can not be used to measure 120 volts AC; it will be destroyed.

Insert the leads from the 120 volt light into the slots, and if the light glows, voltage is present.

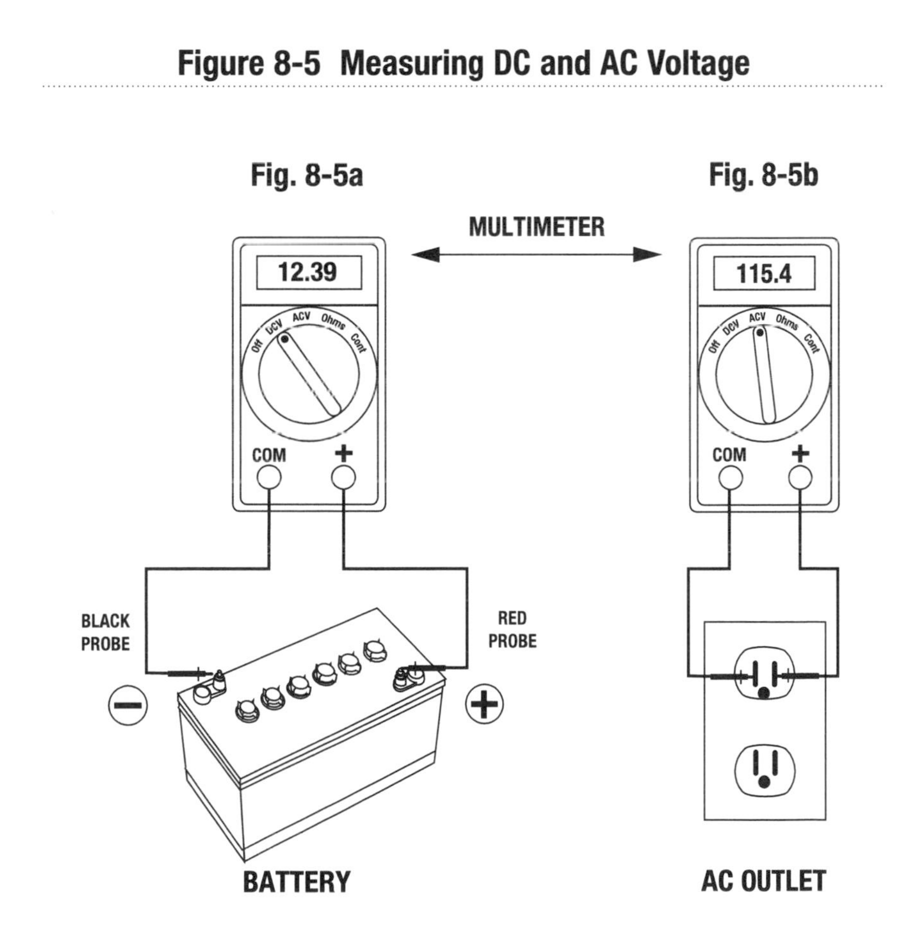

Figure 8-5 Measuring DC and AC Voltage

Measure Resistance

Before measuring a component's resistance, remove all voltage from the circuit and isolate the component from the rest of the circuit. The multimeter's internal battery supplies the current for the resistance reading and the circuit voltage could damage the multimeter. If you do not isolate the component, you may be measuring the circuit resistance and not just the component's resistance. Remove the positive and negative wires from a radio, a stereo, and switches. Remove a fuse or a light bulb from its holder or socket.

Insure that there is no corrosion or oxidation on the component, or the multimeter could indicate a false reading.

Practice Using the Multimeter to Measure Resistance

1. Set the control to the ohmmeter function, either by the function switch or the range selector switch.

2. Touch the two probes together. The reading should be zero or nearly zero; there is no resistance between the two probes. A zero reading indicates the component's resistance is so small the multimeter can not read it.

3. Hold the probes apart, so they are not touching. An unlimited or an infinite amount of resistance exists between the probes. A path does not exist for the electrons to pass, so a break or open circuit condition occurs. The multimeter reading when the probes are held apart is the same reading it shows when an infinite amount of resistance or an open circuit occurs in a component. Some digital multimeters indicate infinite resistance as an OL on the readout; others just go blank. Analog multimeters indicate an infinite amount of ohms when the needle is all the way to the left. What does your meter indicate when you hold the two probes apart? Remember this reading: when measuring resistance or continuity your meter indicates this reading when a break or open circuit has occurred in the circuit or in a component.

Measure a Light Bulb's Resistance

1. Set the control to the ohmmeter function, either by the function switch or the range selector switch, selecting the lowest resistance range if the meter does not have auto ranging.

2. Remove a light bulb from its socket.

3. Place one probe on the tip at the bottom of the bulb and the other probe on the side of the bulb. If the bulb has two tips at the bottom and only one filament, place a probe on each tip. For bulbs with two filaments, place one probe on one tip and the other on the base of the bulb; see figure 8-6A.

A good bulb will indicate a few ohms of resistance. This is not an accurate reading of the bulb resistance because the light bulb's resistance increases as it heats; the bulb's true resistance is when the bulb is hot.

A bad light bulb shows a very high or infinite amount of resistance. If the filament is broken, an open circuit condition has occurred. An infinite or unlimited amount of resistance exists between the two contacts, so electrons can not flow through the circuit.

Fluorescent lights use gas and not a resistive filament, so a multimeter can not be used to determine if they are good.

Measure a Fuse's Resistance

1. Set the control to the ohmmeter function, either by the function switch or the range selector switch, selecting the lowest resistance range if the meter does not have auto ranging.

2. Remove a fuse from its holder.

3. Place a probe on each contact of the fuse; see figure 8-6B. A multimeter indicates a good fuse's resistance is zero or a fraction of an ohm. If the reading is infinite, indicating a break, replace the fuse.

Measure a Radio, Stereo, or Electric Fan's Resistance

1. Set the control to the ohmmeter function, either by the function switch or the range selector switch, selecting the lowest resistance range if the meter does not have auto ranging.

2. Remove the two wires leading to the radio or fan from the power source.

3. Place a probe on each wire. Insure that the device is turned on. If the reading indicates infinite resistance, reverse the probes. If the reading indicates resistance, the device is okay. These types of electrical devices have resistance, and depending on the device,

Figure 8-6 Measuring Resistance and Continuity

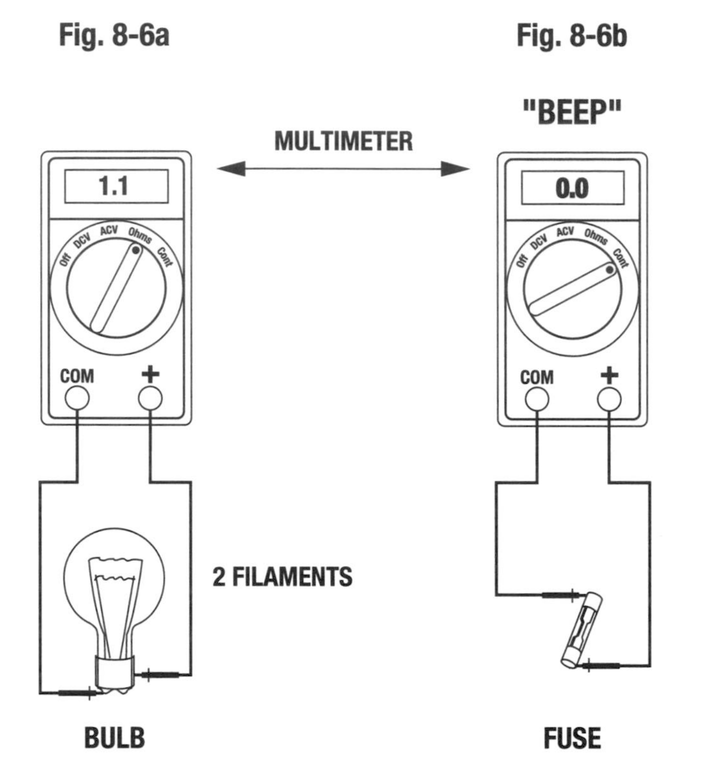

their resistance will range from a few ohms to a couple of hundred ohms. To protect some electronic devices from reverse polarity, diodes are built into the unit. This is the reason resistance is indicated when the probes touch the leads in one position, but not when the probes are reversed.

If the device has zero ohms, there is an internal short.

If an infinite amount of resistance is indicated, there is an open. In both situations, the device is bad and needs to be replaced, or have a technician check the device.

Measure a Switch's Resistance

1. Set the control to the ohmmeter function, either by the function switch or the range selector switch, selecting the lowest resistance range if the meter does not have auto ranging.

2. Electrically isolate a switch from a circuit by removing the wires leading to the switch.

3. Place a probe on each electrical contact of the switch.

4. Turn the switch "on." The multimeter should indicate zero resistance when the switch is in the "on" position. Electricity has a path to flow through the switch.

5. Turn the switch "off." Infinite resistance should be indicated: the switch is breaking the circuit.

CONTINUITY CHECK

A continuity check indicates if components are electrically connected. Some digital multimeters, but not all, indicate continuity by an audible sound. Analog meters indicate continuity when the needle swings all the way to the right, indicating zero or only a few ohms of resistance. When the multimeter is set to measure continuity, do not connect the test leads to a voltage source or damage to the multimeter might result.

1. Set the multimeter to "continuity test" and touch the tips of the probes together. An audible "beep" is heard from a digital meter; the needle of an analog meter indicates zero ohms.

2. Test a fuse or light bulb for continuity. A continuity check determines if a fuse's metal strip or a light's filament has failed. Touch the probes to the contracts of the component, as in figure 8-6. A path exists for electrons if a "beep" is heard from a digital meter, or the analog meter indicates zero or a few ohms of resistance. If there is no continuity, the component is bad and needs to be replaced. The readout indicates if any resistance is present in the component being checked. The multimeter indicates a light bulb's resistance is a few ohms, and the fuse's resistance is zero.

3. Test a switch for continuity when the switch is "on" and for no continuity when the switch is "off."

COMPONENTS OF AN ELECTRICAL CIRCUIT

You have learned about electricity and used a multimeter to measure the voltage of a battery, the resistance of a light bulb, fuse, and radio. Now you need to learn how electricity reacts in electrical circuits.

Every electrical circuit consists of four basic components:

1) **An electrical power source.** A 12 volt battery or 120 volts AC supplied by a municipal utility is an example of an electrical power source.

2) **The load.** The load is an electrical device such as a light bulb, a pump motor, or a radio.

3) **The wiring or conductive path.** The wiring connects the power source to the load. In a flashlight, the conductive path consists of the metal strip in the case, the spring at the bottom and the metal reflector at the top. On RVs and boats, the conductive path is the wiring running throughout the vehicle to connect the battery to the various lights, pumps, and electronic equipment.

4) **The controls.** The switches and fuses in the circuit are the controls that make the electrical energy convenient and safe. A control can be a simple on and off switch or a complicated internal switch in a battery charger that cycles the charger on and off depending on the battery voltage. Fuses and circuit breakers are controls that protect the circuit by breaking the circuit when a short circuit or overload of current occurs.

Each one of these components must be present in an electrical circuit; none is more important than the others. The power source forces the electrons through the circuit, but it's the load that does the work and determines the amount of electrical energy needed. A conductive path is required to move the electrical energy from the power source to the load. Without the switch, the flow would be continuous. Without a fuse, a short will cause an overload of current, destroying the circuit components. They are all important. Without any one of them, the electrical circuit fails.

Figure 8-7 shows a simple electrical circuit with a battery connected by wires to a light bulb. The controls are a simple on and off switch and a fuse.

Figure 8-7 Four Components of a Circuit

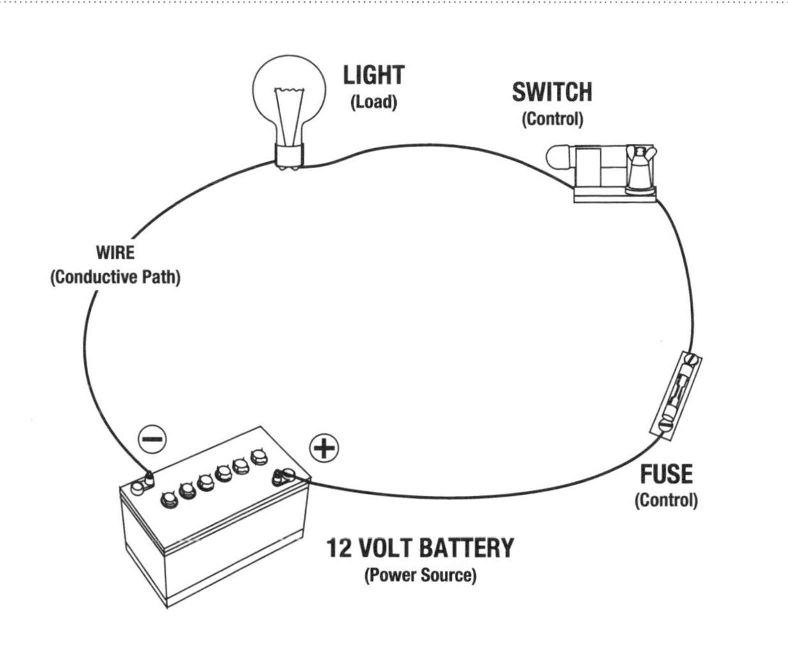

SERIES CIRCUITS

One of the noticeable things about the circuit in figure 8-7 is that the components are connected end to end. The battery negative terminal (the power source) is connected to the negative side of the load by a wire (the conductive path). Another wire attached to the positive side of the load connects the load to the controls (the switch and fuse) and then to the positive side of the battery. This is an electrical circuit connected in series.

A CIRCUIT IS A CIRCLE

It is also apparent that the components are connected as in a circle. In fact, the origin of the word circuit is derived from the word circle. Just as a circle is continuous or endless, a circuit is also continuous. You can break a circle at any point, and it is no longer a circle. You can also break an electrical circuit at any point, and it is no longer a circuit. Normally a circuit is broken or turned off using the switch, but the electrical circuit fails if it is broken at any point. It is sometimes difficult to troubleshoot an open circuit because the break point is not always obvious.

PARALLEL CIRCUITS

The series circuit has only one 12 volt load in the circuit. All RVs and boats have many different 12 volt loads powered by the same 12 volt battery, so they are connected in a parallel circuit.

A parallel circuit consists of two or more loads connected to the same power source. Figure 8-8 shows a circuit with a light and a radio connected in parallel with the battery. Each load is on a separate branch of the circuit.

Each branch of the parallel circuit has the four components of a circuit, and each branch of the circuit is continuous like a circle.

Figure 8-8 Two Branch Parallel Circuit

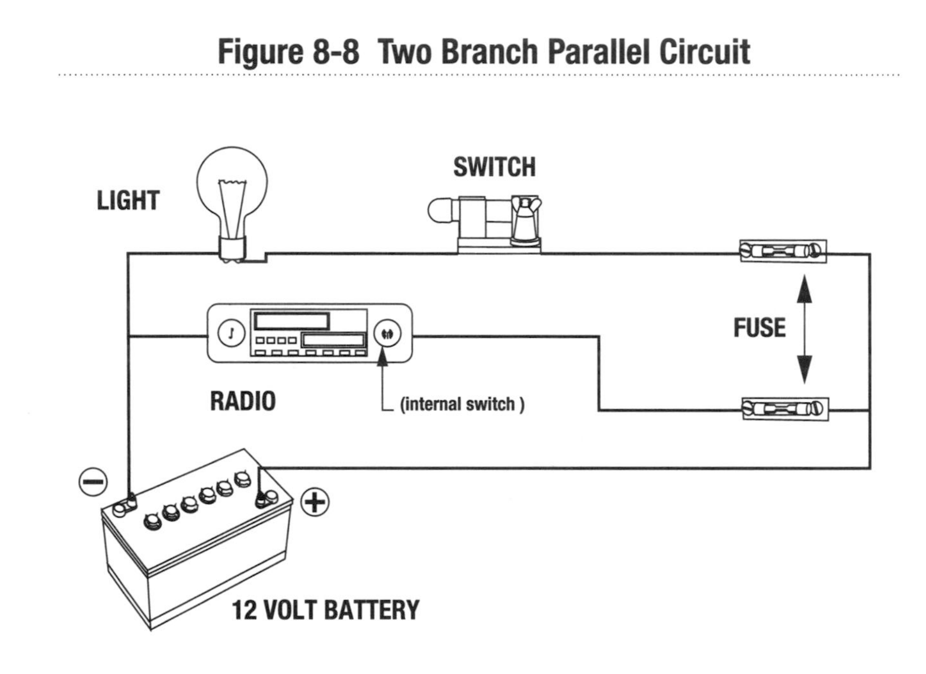

NEGATIVE GROUND

All of the circuits on an RV or a boat have the four essential components of a circuit and are continuous as in the previous examples of series and parallel circuits. One of the first things you may find different when comparing the above illustrations to the circuits in your vehicle is that a few of the negative wires appear to be missing. On an RV, the negative cable from the battery leads to a bolt on the RV's chassis, and on a boat, the cable leads to the engine or to a grounding strap. The negative side of the circuit is not missing for if it were, the circuit would fail. A negative ground system completes the circuit and replaces the wiring for much of the negative side of the circuit. This system reduces costs by decreasing the need to run numerous negative wires back to the terminal of the battery.

Figure 8-9 shows the battery negative cable connected to the grounding system of the vehicle instead of connecting directly to the load. The

Figure 8-9 Ground System and Distribution Panel

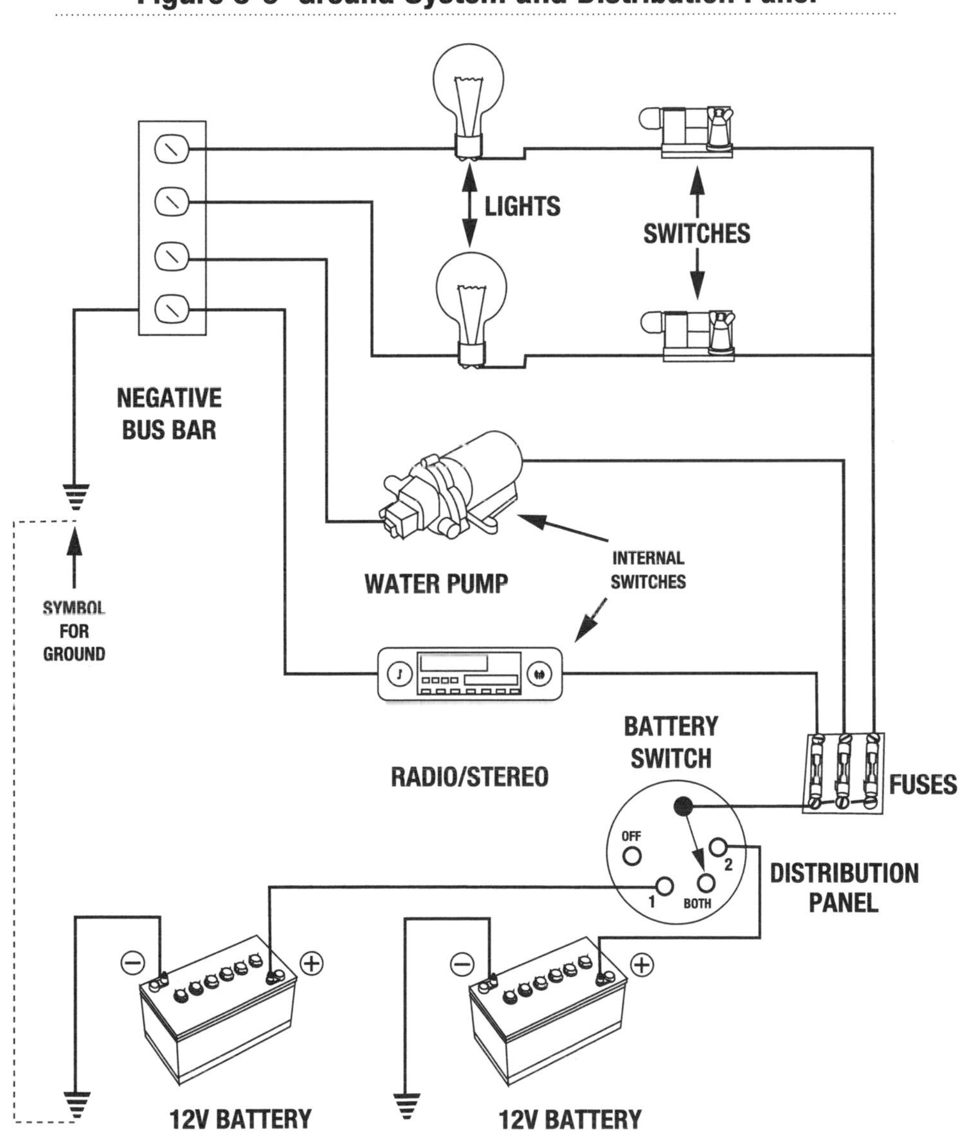

load's negative wires connect directly to the grounding system or to a negative bus bar, a metal bar where several wires can be attached. One wire from the bus bar is then attached to the grounding system, completing the negative side of the circuit.

DISTRIBUTION PANELS

Vehicles have a large number of parallel circuits. Instead of running all the positive wires to the battery positive terminal, one wire from the battery leads to a distribution panel or fuse panel. Figure 8-9 shows some common electrical devices found on an RV or a boat.

The positive battery cable leads to a distribution panel. A distribution panel can be of many types. In a car, the distribution panel is the fuse block containing all the fuses for the lights, horn, radio, heater fan, etc. Some vehicles have fuse blocks that are labeled; others are not labeled, and it is a mystery what they control. Some RVs and boats have impressive distribution panels where labeled circuit breakers replace fuses. The circuit breaker doubles as an on and off switch for each branch of the parallel circuit.

Normally each fuse or circuit breaker supports one load. In figure 8-9, the water pump and the stereo have their own fuse and branch, but the two lights share one fuse and this branch of the parallel circuit branches out to other branches. This is not unusual for lights.

Caution:

In some RVs, the 12 volt DC distribution panel, the 120 volt AC distribution panel and the converter/battery charger are all in the same unit. **A 120 volt AC circuit can KILL.** Before opening up the front panel exposing the wires, unplug the vehicle from the AC outlet and turn off the generator.

IDENTIFYING CIRCUIT COMPONENTS ON YOUR VEHICLE

You must be able to identify the circuit components of your vehicle. If you do not, you will be unable to understand and troubleshoot them. Every vehicle is different, but each 12 volt circuit on the vehicle has the four basic components we have discussed. Just like the simple series circuit in figure 8-7 and the parallel circuit in figure 8-8, the electrical circuits on your vehicle have a power source (12 volt battery), loads (radio or lights), controls (switches and fuses), and conductive paths (wiring and a grounding system).

At your vehicle, try to trace a circuit and identify its components using the following as a guide. Draw the circuit as you trace it, and then you will have a circuit drawing, a schematic, for future use. Your vehicle circuits will look similar to the drawing in figure 8-9.

Power Source

The place to start is at the batteries. A motorized vehicle has a starting battery; locate the starting battery. On RVs, the starting battery is under the hood, and on boats, it is near the engine compartment.

Locate the battery negative terminal and trace the negative battery cable to determine where it is connected. On an RV, the cable bolts to the chassis or a metal object. On a boat, the negative cable bolts to the engine or to a grounding strap. This is the negative grounding system of the vehicle. All the negative wires from the loads must be attached to this grounding system to complete the circuit.

Locate the battery positive terminal and trace the positive battery cable to see where it leads. Tracing this cable can be more difficult because the cable can disappear into the next compartment or lead to a connection were many wires are attached.

On RVs, the battery positive cable connects to other wires. The starting battery supplies power to the engine and to the electrical circuits used in the vehicle operation. You will find one positive wire that runs to the fuse panel that contains the fuses for the horn, radio, lights, etc.

Another wire leads to a solenoid or isolator, which is usually in the circuit to prevent the RV's house circuits from discharging the starting battery. Locate this solenoid or isolator.

On a boat, the positive cable leads to a battery selector switch, so the boater has to select the battery used to start the engine and then isolate the starting battery, preventing it from being discharged.

Once you have located the battery selector switch, solenoid, or isolator, follow the wire from it to the distribution panel. Again it may be difficult to trace the path of this wire because it can disappear under the RV and into nearby compartments. If you can not trace the wire, don't worry, just know it leads to the distribution panel or to the house batteries.

Locate the vehicle house batteries. They are usually located in a different compartment than the starting battery. Trace the negative cables to determine where they are connected. Are they attached to the same grounding system as the starting battery? Sometimes the negative cable leads to a negative terminal on the distribution panel.

Trace the positive cable for the house batteries. Do they lead to a battery selector switch or directly to the distribution panel?

Distribution Panel

The distribution panel contains some of the controls of the circuit. The fuses or circuit breakers located here protect the circuit from excessive current, due mainly to short circuits. If the panel uses circuit breakers, they can also be used as a switch to turn off the circuit.

Distribution panels come in all shapes, sizes, and degrees of sophistication; however, they are all basically the same. The positive wire or cable from the batteries leads to the distribution panel; see figure 8-9. From the distribution panel, the various branches of the parallel circuit branch out to the loads. Each branch has its own fuse or circuit breaker and can support different amperages, so the fuse or circuit breaker can be of different amperage rating. A main circuit breaker may be installed that shuts down all the circuits. Each fuse or circuit breaker should be labeled.

Electrical Loads

The loads are the electrical devices that do the work: lights, pumps, and radios. Check some of the loads on your vehicle. Pull the cover off a light and locate the two electrical wires to the light socket. One is usually red indicating it is positive and the other black, indicating it is returning to ground. The control or light switch is on the positive side of the circuit, so you know the switch is between the distribution panel and the light.

Check other loads such as a radio, a stereo, an electric fan or a pump to see how the electrical wiring is connected to these loads. Where is the control or switch for these units? Radio and stereo controls are built into the unit. Fresh water pumps have pressure sensitive switches that turn on the pumps when the water faucet is turned on. Some electronic devices have a built-in fuse—can you find one on the radio or stereo?

Can you find the fuse or circuit breaker on the distribution panel for the various loads? Several lights may be on one fuse while the other loads have their own fuses or circuit breakers.

Conductive Paths or Wiring

The wiring on an RV or a boat is difficult to trace because it is installed out of sight. The positive wiring path should be easy to visualize. It follows a basic path starting at the battery positive terminal and goes to a selector switch, solenoid, isolator, or directly to a distribution panel.

After the fuse or circuit breaker at the distribution panel, the positive path goes to a switch at or near the load and then to the electrical load.

The negative path is simple; it is from the negative side of the load to ground. The problem is locating the load's grounding point; the grounding point can be anywhere. In figure 8-9, the ground points for the loads lead to a negative bus bar, and then one wire leads to ground. Locating a broken or corroded wire on the negative side of the circuit can be difficult, if the grounding point or negative bus bar is hidden.

PRACTICE WITH A TEST CIRCUIT

The following sections describe how electricity reacts in circuits. It is important to practice using a multimeter or a test light on a known good circuit, so you can understand how electricity reacts. You can build a test circuit like the one illustrated in figure 8-7 using a battery, switch, light bulb and socket, fuse holder and fuse, and lengths of wire. Or, better yet, use a circuit in your vehicle. Pick a circuit that is easy to access. It could be a light circuit where the electrical connections to the light socket are easily probed.

Resistance Determines the Current in a 12 Volt Circuit

High resistance produces low current or amperage.

The resistance of a light bulb filament is about 8 ohms. Using Ohm's law, determine the current drawn by a light bulb when a battery is supplying 12.5 volts.

$$\text{I (current)} = \frac{\text{V (voltage)}}{\text{R (resistance)}} = \frac{\text{12.5 volts}}{\text{8 ohms}} = \text{1.56 amps}$$

Low resistance produces high current or amperage.

A starter motor's resistance is about 0.048 ohms. Again using Ohm's law, determine the current drawn by the starter.

$$\text{I (current)} = \frac{\text{V (voltage)}}{\text{R (resistance)}} = \frac{\text{12.5 volts}}{\text{0.048 ohms}} = \text{260 amps}$$

In both cases, the circuit voltage is the same but because the starter motor resistance is low, the current to the starter is much greater than the current to the light bulb. The wiring, connectors, and controls in the starting motor circuit must be larger than the components in the light circuit because the current is greater.

Determine Power at a Load

Power is the rate an electrical load does work. How much power or watts does it take to light a light bulb compared to the power to start an engine? Power in an electrical circuit is determined by the circuit voltage multiplied by the circuit current. Or simply:

P = V x I.

The power, in watts, to light a light bulb is:

P (watts) = V (voltage) x I (current) = 12.5 volts x 1.56 amps = 19.5 watts

The power to turn a starting motor is:

P (watts) = V (voltage) x I (current) = 12.5 volts x 260 amps = 3,250 watts

The voltage is the same for each load but since the current is different, the watts or power to each load is different. You would expect the power to a starting motor to be much greater than the power to a light bulb because the starting motor requires a greater force to start an engine.

Current Measurement

Measuring the current in a circuit is useful in determining the daily electrical requirement, the amperage output of a charging device, and the amperage drawn by the electrical loads. Also, measuring circuit current can uncover electrical leaks. Some multimeters are able to measure AC and DC current to about 10 amps. Read your meter's instruction manual to insure that it can measure current and note the amount it can measure. Damage to the multimeter occurs if the amperage is greater than the meter is designed to carry. To measure the amperage, the meter is placed in series with the other components, so the circuit amperage flows through the meter. Ammeters are usually placed on the positive side of the circuit, but since the current is the same at any point in the circuit, ammeters can also be installed on the circuit negative side.

Practice Measuring a Light Circuit's Amperage:

1. Read the multimeter's instruction manual to insure that the meter can measure amperage and note the rated current level. Also, the positioning of the probes in the jacks may change over a range of currents.

2. Set the multimeter to measure amperage.

3. Break the circuit, see figure 8-10, at one point and place the probes in line between the power source and the load to measure the circuit amperage or current. Remove a fuse if it is in clips and press the probes to each side of the fuse holder. If you are using an analog multimeter, touch the positive or red probe to a point leading to the positive terminal of the battery, and touch the negative or black probe to a point leading to the negative side of the circuit. This is not required with a digital multimeter. The digital meter indicates a negative sign before the reading if the probes are reversed. Never connect the probes across a power source (between the positive and negative terminals of the battery) when the multimeter is set to read current. The multimeter can be damaged or blow an internal fuse if the current is greater than the current rating for the multimeter.

4. Turn on the circuit; the meter will indicate the circuit amperage.

Figure 8-10 Measuring the Current in a Circuit

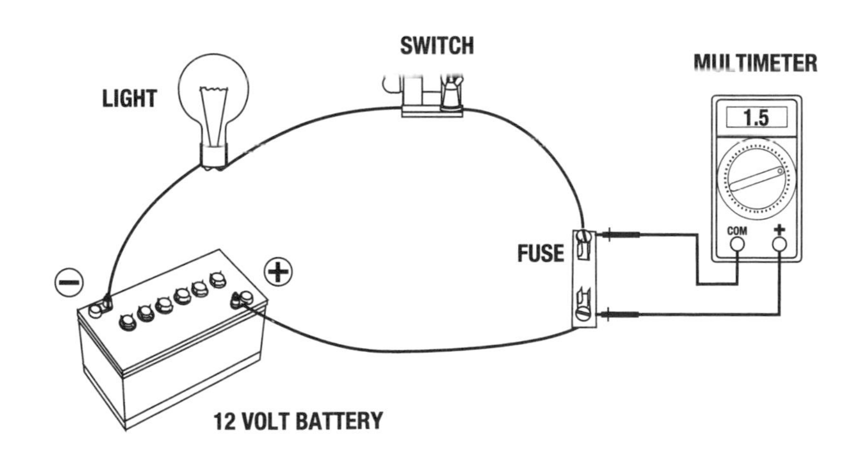

Unwanted Resistance

The electrical load resistance determines the amount of current in a circuit. The load, however, is not the only component that can introduce resistance into a circuit. Long runs of wire introduce resistance due to the wire length and diameter. Corrosion or a loose connection at switches or terminals also causes unwanted resistance.

This unwanted resistance can adversely affect the performance of the load. This is especially true when high amperage is required by a load such as a starter motor.

Ohm's law is used to show the effect a corroded connection has on the current in a light circuit and in a starting circuit. In this example, a corroded connection has a resistance of 0.25 ohms.

Add the light's resistance, 8 ohms, to the resistance of the corroded connection, 0.25 ohms, resulting in a total circuit resistance of 8.25 ohms.

$$\text{I (current)} = \frac{\text{V (voltage)}}{\text{R (resistance)}} = \frac{\text{12.5 volts}}{\text{8.25 ohms}} = \text{1.515 amps}$$

In the starting circuit, the starting motor resistance, 0.048 ohms, is added to the resistance of the corroded connection, 0.25 ohms, resulting in a total circuit resistance of 0.298 ohms.

$$\text{I (current)} = \frac{\text{V (voltage)}}{\text{R (resistance)}} = \frac{\text{12.5 volts}}{\text{0.298 ohms}} = \text{42 amps}$$

The corroded connection has almost no effect on the light circuit, but corrosion reduced the amperage in the starting circuit from 260 amps to 42 amps. (Compare to the example on page 141.) When a large amount of current is required, the circuit resistance must be kept low. In a starting circuit, all the connections and wiring must be free from corrosion and tight because even a fraction of an ohm can mean the difference between successfully starting an engine and failure.

In the following example, you will learn how a corroded connection, with a resistance of 0.25 ohms, affects the voltage of a light and a starting circuit.

The light circuit

Previously, it was determined that the current was reduced to 1.515 amps because of corrosion. The voltage at the light bulb and the corroded connection is also determined using Ohm's law.

Determine the voltage across the corroded connection:

V (voltage) = I (current) x R (resistance) = 1.515 amps x 0.25 ohms = 0.38 volts

Voltage at the light:

V (voltage) = I (current) x R (resistance) = 1.515 amps x 8 ohms = 12.12 volts

The corroded connection is producing a voltage drop of 0.38 volts; see figure 8-11. The total circuit voltage is equal to the battery voltage, 12.5 volts (12.12 volts at the light plus 0.38 volt at the corroded connection), but because of the voltage drop at the connection, the voltage received by the load or light bulb is reduced. The light is dimmer, but it still works.

Figure 8-11 Voltage Drop in a Circuit

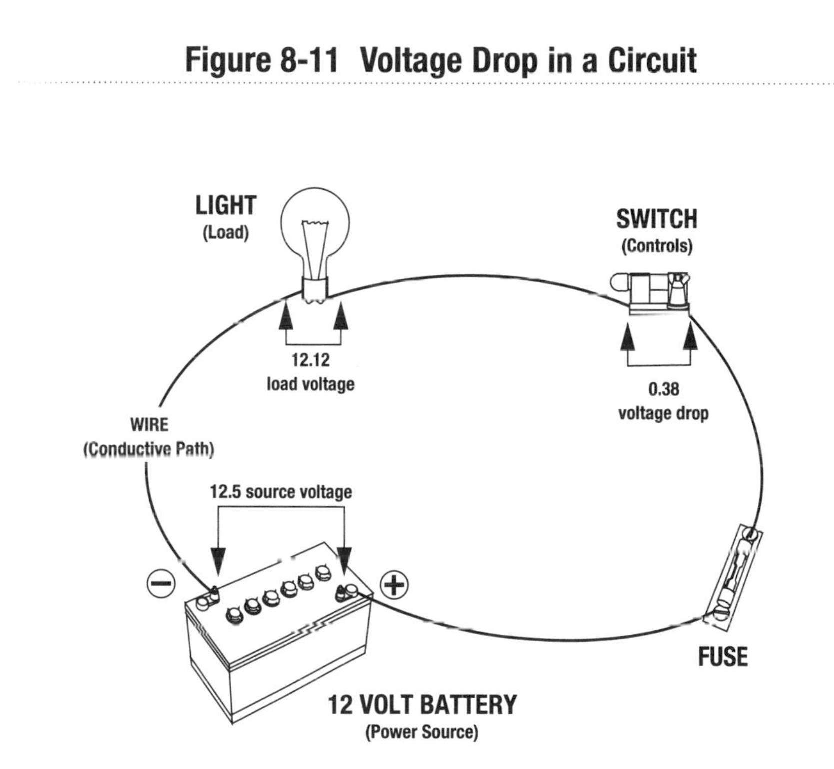

The Starting Circuit

Previously, it was determined that the starting circuit current is reduced to 42 amps because of corrosion.

Again, the voltage of the starter and the corroded connection is determined using Ohm's law.

Determine the voltage drop at the corroded connection.

V (voltage) = I (current) x R (resistance) = 42 amps x 0.25 ohms = 10.5 volts

Determine the voltage at the starter:

V (voltage) = I (current) x R (resistance) = 42 amps x 0.048 ohms = 2 volts

The voltage drop at the corroded connection is now 10.5 volts because the amperage is greater in the starting circuit than the amperage of the lighting circuit. The starter motor is not receiving enough voltage to start the engine.

Both examples have 12.5 volts in the circuit, but where a high amount of current is required by the load, unwanted resistance has a devastating effect on the performance of the circuit. Unwanted resistance could be one reason why a circuit does not perform properly.

The corroded connection in the starting circuit becomes extremely hot because it dissipates the voltage as heat. Excessive heating of a component makes this an easy way to find unwanted resistance or a voltage drop. If a starting circuit fails to start an engine, feel the battery terminals and connections for hotter than normal components. The wiring or connections in any circuit should not be hot. If they are, they are a source of unwanted resistance and perhaps the reason the load is not performing properly.

All wiring has resistance—the smaller the wire, the longer the run, the greater the resistance. Some of the metals used in wiring are more conductive than others. Silver is very conductive but impractical. Copper is slightly less conductive than silver and should be used in RVs and boats. The resistance inherent in wiring causes a voltage drop in circuits; a three percent drop is not uncommon. In a 12.5 volt circuit, a 3 percent drop amounts to a voltage drop of 0.4 volt for long runs of wiring. To minimize the voltage drop due to the resistance of wires, keep wiring runs short and use the proper size wire for the amount of current the circuit will carry.

Measuring a Circuit's Voltage

Using a multimeter measure the voltage at a load.

1. Set the multimeter on DC volts.

2. Turn on the circuit to be tested so the load is operating: the light is lit, or the radio is playing.

3. Place a probe on each battery terminal to measure the voltage source. This is the battery closed circuit voltage and is less than the battery open circuit voltage. The greater the circuit current, the greater the difference between the battery open circuit and closed circuit voltage.

4. Place a probe on each connection at the load to measure the load voltage; see figure 8-12. (It can be difficult to find a load where you can access the positive and negative wires while the load is on.) The voltages should be the same. If they are not, a voltage drop is occurring. Long runs of wire reduce the voltage at the load by 3 percent or by 0.4 volts, which reduces the voltage at the load to 12.1 volts. A voltage drop is caused by resistance in the circuit components. Corrosion, loose connections, long wiring runs, and undersize wiring introduce unwanted resistance into an electrical circuit.

Figure 8-12 Measuring a Circuit's Voltage

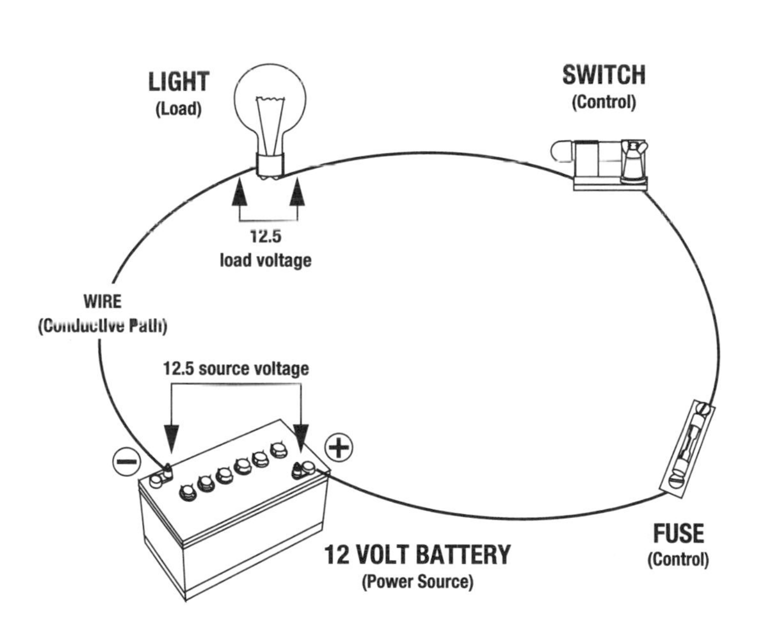

Finding a Voltage Drop

A voltage drop in a simple circuit, as in our examples, may be difficult to detect. Fuses have a voltage drop of a few hundredths or tenths of a volt, and the fuses at distribution panels are usually easy to probe, so try to measure their voltage drops. You can also try to measure voltage drops at other circuit components. (Figure 8-11 is an example where a voltage drop is measured at a switch.)

1. Turn on the circuit. The circuit the fuse is in must be operating.

2. Place the red probe on the end of the fuse or fuse holder leading to the battery positive terminal. Place the black probe on the other end of the fuse or fuse holder. The voltage drop should not be more than a few tenths or hundredths of a volt.

3. Turn off the circuit. The multimeter will indicate zero volts. A voltage drop only occurs when the circuit is operating.

Finding the Voltage of an Open Circuit

1. Turn on the circuit. The load should operate; the light lights, the radio plays.

2. Measure the voltage at the switch or fuse for this circuit, whichever is easiest to probe. It should measure zero volts. If the switch or fuse contains a significant amount of resistance, it may have a voltage drop of a few tenths of a volt.

3. Turn off this switch or pull this fuse to break the circuit, causing an open circuit. The load ceases to operate. The voltage reading at the switch or fuse is now 12.5 volts.

4. Measure the load voltage. It is zero volts. Figure 8-13 shows what happens in this example. The circuit is broken, and the battery voltage is no longer measured at the load, but at the break in the circuit. **When you find the voltage in an open circuit, you have found the break in the circuit.**

5. Turn off the other control in this circuit; the switch if in step 3 you pulled the fuse. The voltage at the fuse and the switch is now zero volts (your multimeter reading may jump around or indicate a few hundreds or thousands of a volt).

Figure 8-13 Voltage in an Open Circuit

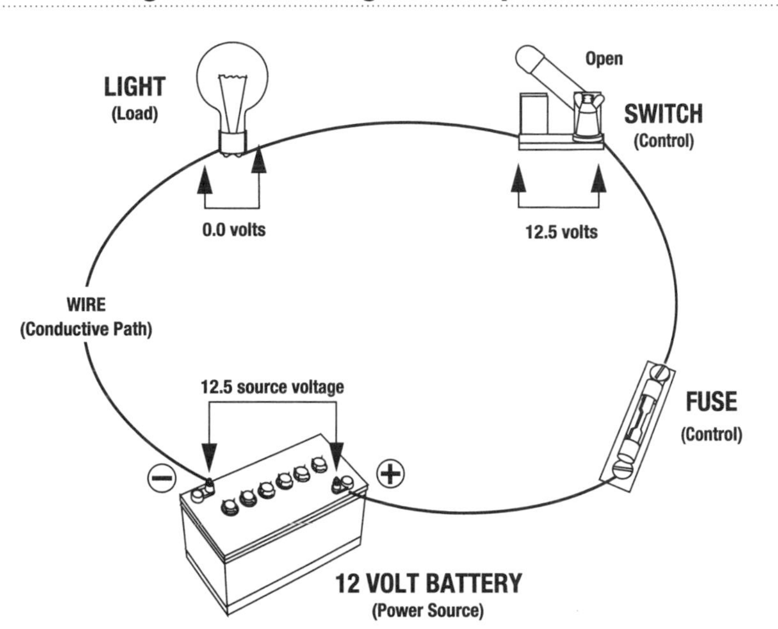

Figure 8-14 Two Breaks in a Circuit

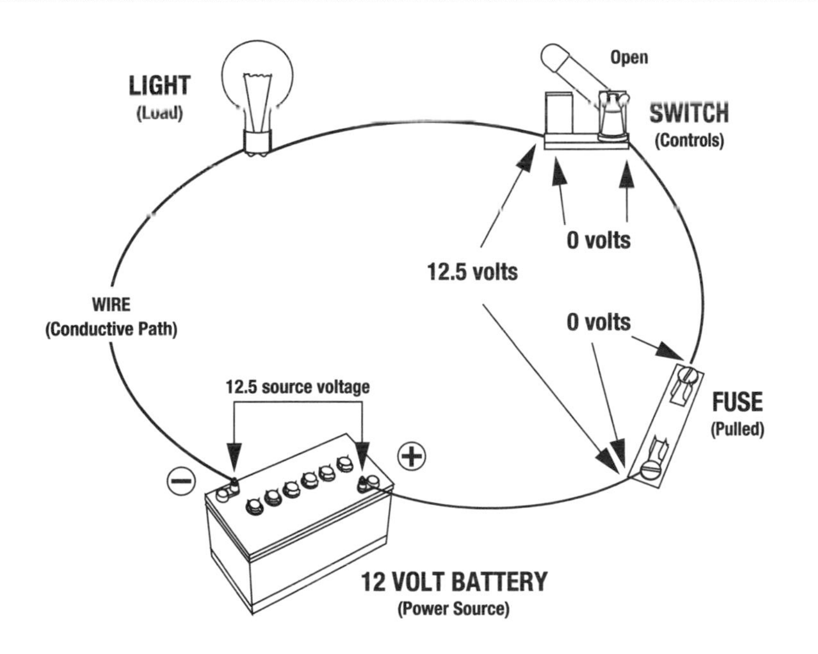

6. Probe the positive side of the fuse and the negative side of the switch; see figure 8-14. The multimeter will read 12.5 volts. The break is now in two places and the voltage is between the switch and fuse. Finding the break could be difficult. Fortunately, a circuit normally fails only at one point.

An open circuit is one reason a circuit fails. In the above example, you have learned that by turning off the switch or pulling the fuse you have broken the circuit and the load turns off. The circuit voltage is now at the break and not at the load. This makes a good method of finding a break in an open circuit because if you find the voltage you have found the break.

PROBLEM WITH THE POWER SOURCE

The battery current and voltage must support the energy requirement of the loads, or the loads will not function properly. Flashlight batteries are replaced when the light of a flashlight dims because the batteries can no longer supply the current and voltage required by the light bulb. When the voltage of a 12 volt battery decreases to a low level, it is recharged.

As the battery voltage decreases during normal discharging, the battery has difficulty supplying a large amount of current, but not a low amount of current. Lead sulfate builds on the battery plates during the normal electrochemical process. This increases the battery internal resistance, and this resistance has the same effect as the other resistances discussed in the previous examples. When the current is kept low, the internal resistance has little effect on the circuit, but for high current requirements the internal resistance may cause the circuit to perform inadequately.

A battery in a high state of charge will have an open circuit voltage of 12.5 volts, and depending on the battery design an internal resistance of only 0.006 ohms.

Determine the current supplied to a 8 ohm light bulb and a 0.048 ohm starter.

Use ohm's law to determine the light bulb's current:

$$\text{I (current)} = \frac{\text{V (voltage)}}{\text{R (resistance)}} = \frac{\text{12.5 volts}}{\text{8 ohms + 0.006 ohms}} = \text{1.56 amps}$$

Again, use Ohm's law to determine the starter motor's current:

$$I\ (\text{current}) = \frac{V\ (\text{voltage})}{R\ (\text{resistance})} = \frac{12.5\ \text{volts}}{0.048\ \text{ohms} + 0.006\ \text{ohms}} = 231\ \text{amps}$$

A voltage drop occurs whenever current encounters resistance. Just like the voltage drop at the corroded connection on page 145, a voltage drop occurs due to the battery internal resistance when the light bulb or the starter motor is turned on.

Ohm's law is used to determine the voltage drop due to the battery internal resistance when the light is turned on.

V (voltage) = I (current) x R (resistance) = 1.56 amps x 0.006 ohms = 0.01 volt

Voltage at the light bulb:

V (voltage) = I (current) x R (resistance) = 1.56 amps x 8 ohms = 12.48 volts

The voltage drop is very small because the light circuit amperage is low.

Ohm's law is again used to determine the voltage drop due to the battery internal resistance when the starter motor is used to start an engine.

V (voltage) = I (current) x R (resistance) = 231 amps x 0.006 ohms = 1.4 volts

Voltage at the starter motor:

V (voltage) = I (current) x R (resistance) = 231 amps x 0.048 ohms = 11.1 volts

Only 11.1 volts is reaching the starter motor because of the battery internal resistance and the starter circuit's higher current compared to the light circuit's current.

The 11.1 volts is the closed circuit voltage of the starter motor circuit. Even through the internal resistance is the same for both examples, the greater the curent in the starter circuit means less voltage supplied to the load. The closed circuit voltage of a circuit is always less than the open circuit voltage and the greater the current the greater the difference.

If a voltmeter is installed in the electrical system, you can observe the voltage drop when a light bulb or a starter motor is turned on. When the light bulb is turned on, the voltage drops by a small amount compared to when a starter is turning over an engine.

The power that the battery can supply to the light and starter motor is determined by multiplying the closed circuit voltage times the current or:

$$P = V \times I$$

The power to the light bulb is:

$$P = V \text{ (voltage)} \times I \text{ (current)} = 12.48 \text{ volts} \times 1.56 \text{ amps} = 19 \text{ watts}$$

The power to the starter motor is:

$$P = V \text{ (voltage)} \times I \text{ (current)} = 11.1 \text{ volts} \times 231 \text{ amps} = 2564 \text{ watts}$$

A battery with a high state of charge has no problem supplying the needed power to both the light bulb and the starter motor.

A deeply discharged battery with an open circuit voltage of 12.0 volts may have an internal resistance of 0.01 ohms. Determine the current supplied to the light bulb and the starter motor.

Ohm's law determines the light bulb's current:

$$I \text{ (current)} = \frac{V \text{ (voltage)}}{R \text{ (resistance)}} = \frac{12.0 \text{ volts}}{8 \text{ ohms} + 0.01 \text{ ohms}} = 1.498 \text{ amps}$$

Again, Ohm's law determines the starter motor's current:

$$I \text{ (current)} = \frac{V \text{ (voltage)}}{R \text{ (resistance)}} = \frac{12.0 \text{ volts}}{0.048 \text{ ohms} + 0.01 \text{ ohms}} = 207 \text{ amps}$$

Determine the voltage drop due to the battery internal resistance when the light is turned on.

$$V \text{ (voltage)} = I \text{ (current)} \times R \text{ (resistance)} = 1.498 \text{ amps} \times 0.01 \text{ ohms} = 0.01 \text{ volts}$$

Voltage at the light bulb:

$$V \text{ (voltage)} = I \text{ (current)} \times R \text{ (resistance)} = 1.498 \text{ amps} \times 8 \text{ ohms} = 11.98 \text{ volts}$$

The voltage drop is very small because the light circuit amperage is low.

Ohm's law is again used to determine the voltage drop due to the battery internal resistance when the starter motor is used to start an engine.

V (voltage) = I (current) x R (resistance) = 207 amps x 0.01 ohms = 2 volts

Voltage at the starter motor:

V (voltage) = I (current) x R (resistance) = 207 amps x 0.048 ohms = 10 volts

Now only 10 volts is supplied to the starter motor.

The power that the battery can supply to the light and starter motor is determined by multiplying the closed circuit voltage times the current or:

P = V x I

The power to the light bulb is:

P = V (voltage) x I (current) = 11.98 volts x 1.498 amps = 18 watts

The power to the starter motor is:

P = V (voltage) x I (current) = 10 volts x 207 amps = 2070 watts

The light bulb is dimmer but is still on, but because the starter motor requires a large amount of current, its power is reduced to 2070 watts from 2564 watts.

As a battery becomes deeply discharged, it has a difficult time supplying sufficient power to starting motors, inverters, or other high current loads, but when the amount of current is reduced this deeply discharged battery can supply power to low amperage loads.

The answer to this problem is simple; do not allow your batteries to become deeply discharged, and properly maintain them.

Chapter 9

Troubleshooting Electrical Circuits

If you understand how to troubleshoot electrical circuits, you will be able you to repair most of your electrical problems.

This chapter explains how to troubleshoot some of the common circuits found on RVs and boats. What is **not** in the scope of this book is an explanation on how to fix some of the more complicated components of the electrical circuit. Radios, alternators, or battery chargers should be taken to a qualified technician for repair, if they are determined to be bad.

This troubleshooting guide should be clear and understandable, but if at anytime you feel you are in over your head or do not understand what is going on, **stop!!** Electricity can **kill.** Sometimes it is better to wait for a technician, and pay him to solve your electrical problems.

SCHEMATICS

Schematics or wiring diagrams are useful tools when trying to troubleshoot a failed circuit. A schematic indicates how the loads are connected to the power source and where the switches and fuses or circuit breakers are located. Unfortunately, most RV and boat manufacturers do not provide detailed schematics of the electrical systems on board a vehicle.

If your vehicle does not have a schematic, it is wise to draw your own. Draw at least the distribution panel and label each fuse, indicating what it controls. This is not difficult to do when all the circuits are functioning properly. Turn on all loads and pull out a fuse to see what turns off. Also, label each fuse at the panel, so when troubleshooting a failed circuit it is easy to locate its fuse.

Add all new equipment to the schematic. Adding to a schematic only takes a few extra minutes, and the information is invaluable if the circuit ever fails. While troubleshooting a failed circuit, you usually locate all the circuit components, so take a few minutes to add the component's location to the schematic. Without a schematic, it takes time to learn the location of all the controls, wiring, and connections in a circuit. If you update your schematic, the next time you have to troubleshoot the circuit it will go much smoother.

The schematic should have the location of the circuit components: power sources; any selector switches, solenoids, or isolators; distribution panels with the fuses labeled and their amperage rating listed; electrical loads; grounding points of the loads; and wiring leading between the components, with color and terminal numbers recorded.

REASONS CIRCUITS FAIL

Electrical circuits fail because of an open circuit, or a short circuit in the load or circuit. Also, voltage drops can rob the load of the necessary voltage required for proper operation.

Find a Break in an Open Circuit

Current does not flow in an open circuit, so the electrical load (light, radio, pump) does not operate. Find the break in the circuit.

1. Insure that all the switches are "on." Know about all controls; several switches could control the circuit. When troubleshooting, if you leave a switch off or don't fix a blown fuse or a tripped circuit breaker, you have caused a second break in the circuit as in figure 8-14. This makes finding the failed component much more difficult.

2. Check the circuit fuses to insure that none is blown. Check circuit breakers to insure they have not tripped; if they have, check for short circuits. See section on "Find a Short Circuit" page 160.

3. Set the multimeter to DC volts.

4. Measure the voltage at the load by placing the red probe to the positive connection to the load and the black probe to the negative or ground connection to the load. (The wires may be disconnected from the load to do this.)

   A. If the voltage is 12 volts, the break is internal to the load. To check for an internal break:

      1. Insure that the internal load switches are "on."

      2. Insure that the internal fuses are not blown.

      3. Set the multimeter to the ohm position.

      4. Disconnect the wires from the load.

      5. Measure the resistance of the load. Most loads have resistance; fluorescent lamps are an exception. If an infinite amount of resistance is measured, the open circuit is caused by the load. If the load is an electronic device, reverse the probes and recheck. Replace or repair the load.

   B. If the voltage is zero at the electrical load, electricity is not reaching the load and the break is elsewhere in the circuit. To check for a break in the circuit:

      1. Set the multimeter to DC volts.

      2. Probe the other circuit components by placing the two probes on either side of the component to be tested. If you measure 12 volts, you have found the break. Repair the cause of the break. Figure 8-13 is an example where voltage is found at a break in the circuit. If you are unable to find the break after testing each component, check for voltage from the load to ground or from the load to the positive battery terminal. This is difficult to do, however, because the probes on the multimeter are short. A jumper wire can be used to help locate the break. Any wire long enough to reach a ground point or a positive point on the distribution panel or the battery terminal will do. Alligator clips are useful in attaching the jumper wire to a connection.

a. Check the ground side of the load. Attach one end of the jumper wire to a grounding point on the vehicle and connect the other end of the jumper wire to the black probe of the multimeter; see figure 9-1. Touch the multimeter's red probe to the negative connection of the load. Insure that oxidation or corrosion is not on the connections for it will give you a false reading; scrape the connection until bright metal appears. If the voltage is 12 volts, the problem is a poor ground. Find the cause and repair. If the voltage is zero, the problem is not on the ground side of the circuit.

b. Check the positive side of the circuit. Attach one end of the jumper wire to the positive battery terminal or to a known good positive terminal. Connect or touch the other end to the multimeter's red probe. Touch the black probe to the positive connection of the load. If the voltage is 12 volts, the break is on the positive side of the circuit. Use the jumper wire to narrow the hunt for the break by eliminating sections of the circuit. Leave the jumper wire connected as described above. Touch the multimeter's black probe to the most accessible terminals first: switches, terminal blocks, or connections. If the multimeter indicates 12 volts, the break is from that point to the battery positive terminal or to where you connected the jumper wire. If the voltage is zero, the problem is from that point to the load.

As an example, see figure 9-1. The break in this circuit is a broken wire at the bus bar. Using a jumper wire and a multimeter, the ground side of the circuit is checked as described above. The multimeter indicates zero volts, so the positive side of the circuit is checked. A jumper wire is attached to the fuse and to the multimeter's red probe. The black probe touching the positive connection at the light or the switch indicates 12 volts on the multimeter. Since the multimeter indicates 12 volts at each component, you know the break point is between the switch and the fuse. By checking the bus bar, you find the broken wire.

Figure 9-1 Finding a Break in an Open Circuit

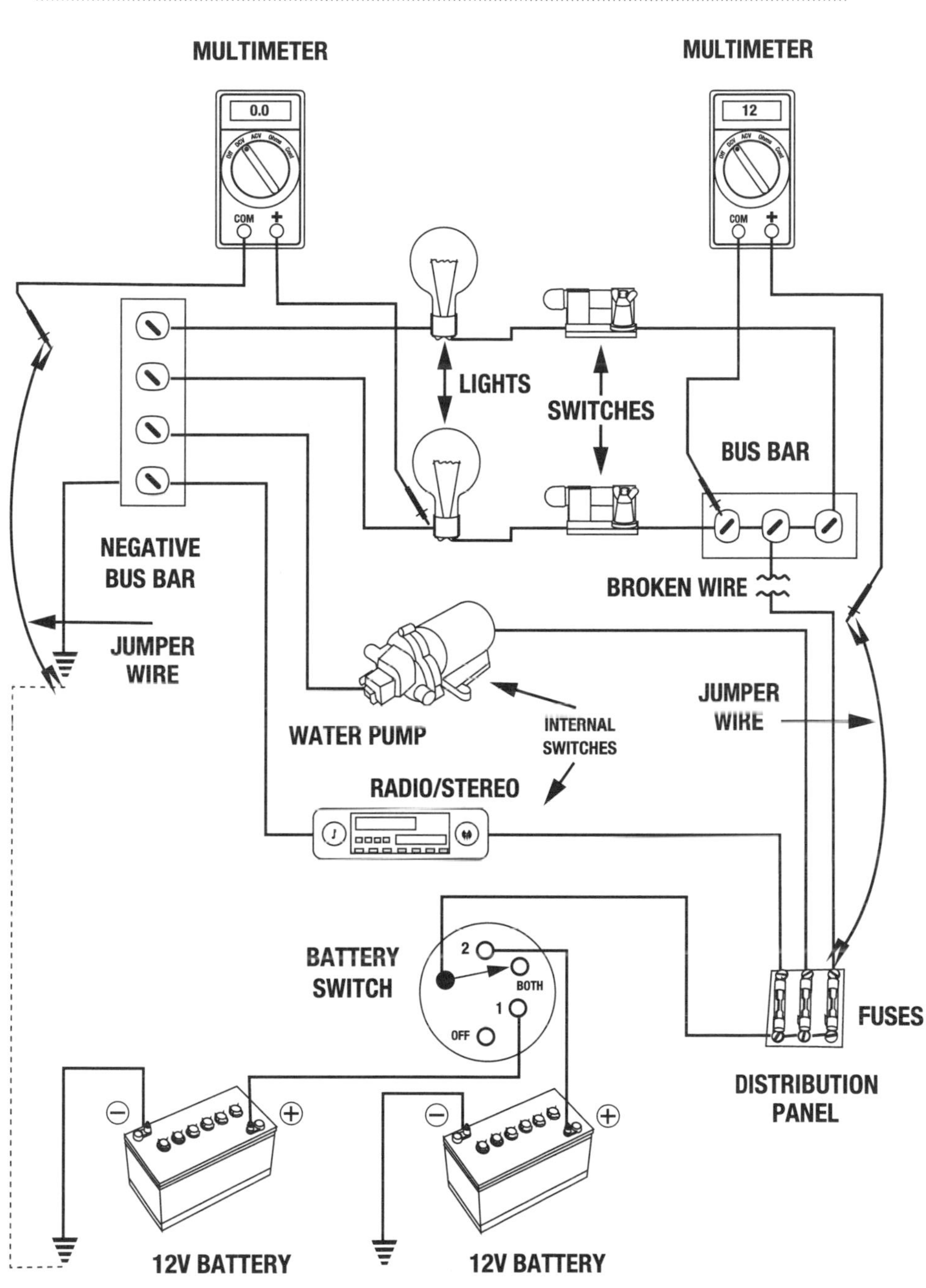

Find a Short Circuit

A fuse fails or circuit breaker trips when the current finds a direct path to ground and bypasses the load. Without the load resistance in the circuit, a large amount of amperage flows through the circuit, blowing fuses, tripping circuit breakers, or melting a component. This is called a short circuit, and its cause must be found and corrected. Excessive loads placed on a circuit will also blow a fuse or trip a circuit breaker; decrease the circuit electrical loads.

To find a short circuit:

1. Check the obvious. Check the circuit for any wires or metal objects that would cause current to flow to ground, bypassing the load. A short could occur because a positive wire is touching a grounding point, or a screw or some metal object has fallen across a terminal, shorting it out. A melted wire may be the result of a short but not the cause. If a fuse fails to blow, the next most sensitive component fails.

2. Replace the fuse, or reset the circuit breaker, if you do not locate a problem. Never replace a fuse with a higher rated fuse—doing so may result in overheating a circuit component and in a fire. If the fuse blows again or the circuit breaker trips, go to step 3. Figure 9-2 shows a short circuit bypassing the load.

3. Set the multimeter to ohms or test for continuity.

4. The short may be internal to the load. Disconnect the positive side of the circuit from the power source, and disconnect the load from the circuit by removing the positive and negative wires. Replace the fuse with a good one and insure that the switch is on. Test the load by touching the multimeter probes to each electrical lead to the load. Most loads have resistance; fluorescent lamps are an exception. Reverse the leads and recheck, if the load is an electronic device. You should expect some resistance reading. If the resistance is zero ohms, the short is internal to the load, and the load needs to be repaired or replaced. Some loads such as starting motors and electrical anchor windlasses have very low resistance. When using an inexpensive multimeter, the meter may indicate zero resistance even when the motors are good. Have the motors tested by a qualified technician.

Figure 9-2 Finding a Short Circuit

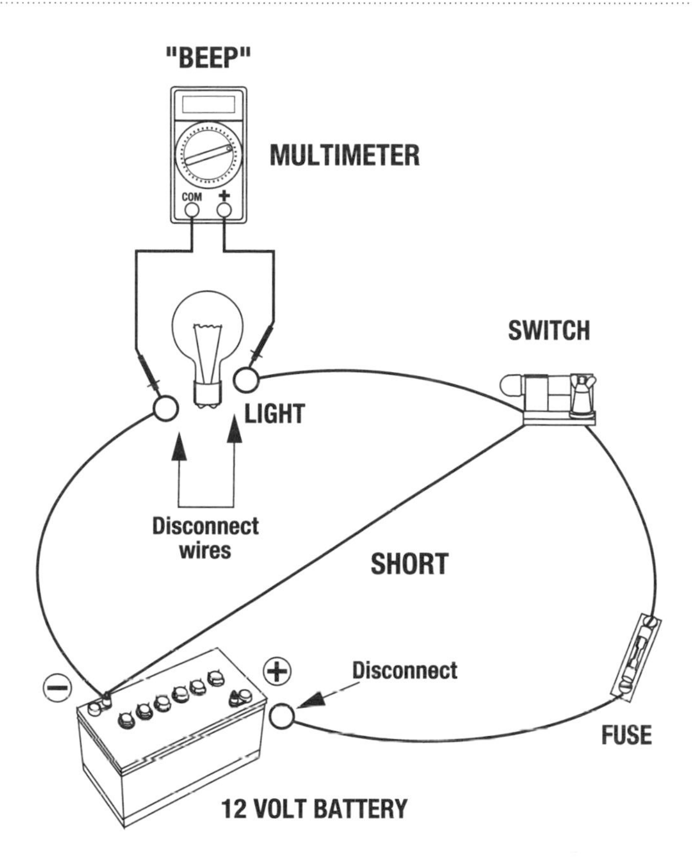

5. If the load is okay, the short is elsewhere in the circuit. To find a short circuit, you check for continuity where it should not be. There should be no continuity between the negative side of the load and the positive side of the load, because the leads to the load have been disconnected. To check, touch the multimeter probes to each wire leading from the load; see figure 9-2. If the reading is zero, a few ohms, or continuity is indicated, a short circuit situation exists, and the cause must be found.

6. Test the positive side of the circuit for a short. Leave the probes attached to the connectors or wires on both sides of the load. Pull out the fuse. If the multimeter now indicates infinite resistance, the short is between the fuse and the positive battery terminal. If the multimeter still indicates continuity or a few ohms, the short is not located on the positive side of the fuse. Replace the fuse.

7. Turn off the switch. If the multimeter indicates infinite resistance, the short is between the fuse and the switch. If the multimeter still indicates continuity, the only place left for the short to occur is between the switch and the load. In this example, that is where the short is.

Find a Voltage Drop

A load may not be operating properly because voltage is being lost at unwanted resistance in the circuit connections, switches, or wiring. Conduct the following test to find an unwanted voltage drop in a switch, fuse, run of wire, or other component. The circuit must be operating; a voltage drop does not occur unless current is flowing through the circuit.

1. Set the multimeter to DC volts.

2. Place a probe on one side of the component to be tested. Place the other probe on the other side of the component. The component to be tested must be between the two probes, such as the switch in figure 8-11.

3. If the multimeter indicates zero volts, a voltage drop does not exist at the component. If the multimeter indicates voltage, a voltage drop exists at the component: the greater the amount of current, the greater the voltage drop.

4. Probe the other components of the circuit, checking for a voltage drop. Long runs of undersized wiring can cause excessive voltage drops. Increasing the wiring size reduces the voltage drop.

5. Depending on the electrical load, a voltage drop greater than 0.5 volts indicates a problem. Also, heat is generated at components with unwanted voltage drops. Check components for excessive heat. Clean, repair, or replace the component to reduce the resistance.

TROUBLESHOOTING COMMON PARALLEL CIRCUITS

The preceding troubleshooting procedures are used on any circuit, but since the circuits found on RVs and boats are the more complicated parallel circuits, the following sections are included to help you troubleshoot these circuits.

Figure 9-3 A Common Parallel Circuit

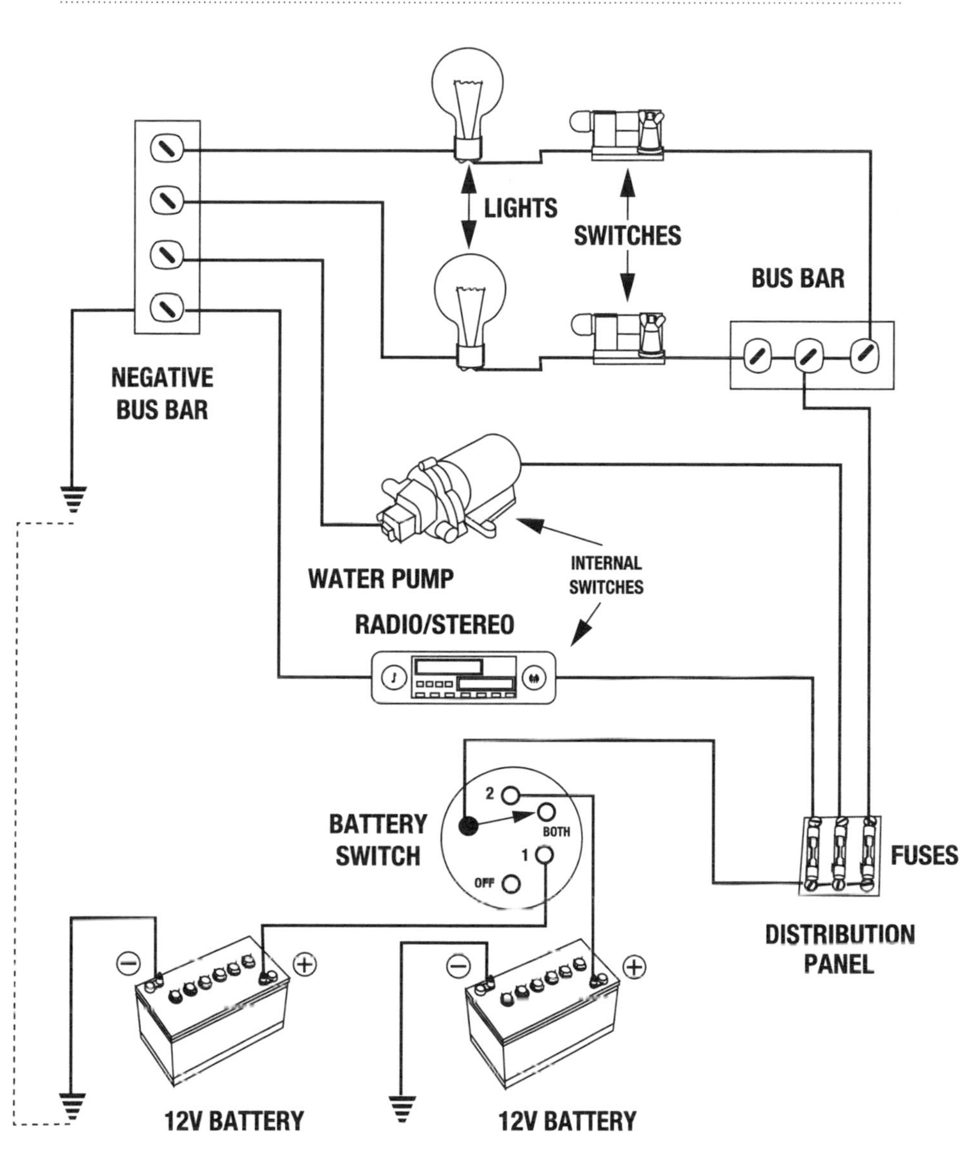

All Electrical Loads Fail to Operate

In figure 9-3 if the lights, pump, and radio fail to come on, it is unlikely that all of them failed, so the problem is a component they have in common. In a parallel circuit, the many branches are powered from a central power source and controlled from a distribution panel that may contain a main power switch. Check these components first to locate the reason for failure.

1. Check the battery selector switch, if one is installed, to insure it is not turned off.

2. Check the main circuit breaker or fuse on the distribution panel.

3. Check the batteries for voltage. If they have no voltage, recharge the batteries. Determine why the batteries lost voltage. See "Determine the Cause of Battery Failure" on page 174.

4. Check the battery terminals and all connections to insure that they are tight and corrosion free.

Failed Lighting Circuits

When a light fixture fails, the most common problems are fuses, bulbs, corroded fittings and loose wires. Check these items first when a lighting circuit fails. If these are not the problem, it is sometimes difficult to locate the failure. It is difficult to trace lighting circuits through a vehicle because the wires spread throughout the vehicle. Many lights have a common fuse but their own switch and grounding points. Some lighting circuits have as many as three switches controlling the circuit: the battery selector switch, the circuit breaker, and the switch at the light. Understand all the controls in a circuit because it only takes one to turn off the circuit.

If all the lights have failed in a lighting circuit but other loads are operating normally, a common component to all the lights has failed. In figure 9-3 if both lights fail, it is unlikely both switches and bulbs are bad.

1. Check the fuse; if it is blown, check for a short circuit. See "Find a Short Circuit" on page 160. If the fuse or circuit breaker is good, look for an open circuit.

2. Check for an open circuit. Follow the procedure as indicated in "Find a Break in an Open Circuit" on page 156. It's unlikely both bulbs and switches failed, so you need to check the common positive and ground sides of the circuit.

Failed Electronic Devices

Radios, stereos, TVs, and other electronic devices usually have their own branch of a parallel circuit. If the other loads, lights or pumps, are operating normally, the problem is in the electronic device's branch of the failed circuit.

1. Check the fuse or circuit breaker at the distribution panel; if it is blown or tripped, check for a short circuit. See "Find a Short Circuit" on page 160.

2. Electronic devices usually have an internal switch and fuse; insure that they are operating properly.

3. Check the electronic device's instruction manual to see if it has a troubleshooting guide.

4. Check for an open circuit. Follow the procedure as indicated in "Find a Break in an Open Circuit" on page 156.

5. Check for a voltage drop. The load may not be receiving the voltage it requires because of a voltage drop or because the batteries are deeply discharged. Follow the procedure in "Find a Voltage Drop" on page 162. Ham and SSB radios require 20 amps when transmitting, so the voltage drop will be greater when the radio is transmitting than when receiving. Check for a voltage drop while the radio is transmitting.

6. Check the voltage at the battery terminals when the ham radio is transmitting. "Problem with the Power Source" on page 150 shows that when a large amount of current is required, a deeply discharged battery may not be able to supply the voltage necessary for the load to operate properly. Keep your batteries charged and in good condition.

TROUBLESHOOTING CHARGING CIRCUITS

A charging circuit fails for the same reasons other electrical circuits fail: open circuits, short circuits, and voltage drops keep some of the voltage from reaching the load. The basic components of a circuit are present, and the circuit is continuous like a circle, but the battery is now the load and not the power source. In a light or radio circuit, the battery is supplying electricity to the electrical load. In a charging circuit, a charging device—alternator, battery charger, or solar panel—is supplying electricity to the battery to recharge it, so the load is the battery.

In a radio or light circuit, the voltage supplied to the radio or light is the battery voltage. In a charging circuit, the voltage of the charging device must be greater than that of the battery. If the charging voltage is the same as the battery voltage, the battery will not be charged. The voltage and current supplied by the charging device determines how quickly and efficiently the battery is charged. The greater the charging voltage and amperage the more quickly the battery is charged, however, they must be controlled as the battery approaches full charge. The charging process is discussed in Chapters 3 and 5.

Check a Charging Device's Output

It is important to measure and monitor the voltage and current supplied to the battery by a charging system. If the voltage and current are high, the battery will be charged quickly. If the voltage and current are low, it takes many hours to recharge the battery.

To check the charging system:

1. Using a multimeter set on DC volts, measure the battery voltage at the terminals while the charging device is on. If the battery voltage is increasing, the charging device is probably working. The battery voltage should increase to at least 12.8 volts. If not, see the troubleshooting section for alternators, battery chargers, solar panels, or wind generators—whichever is appropriate.

2. If the voltage is low (12.8 to 13.2), determine the charging device's rated output. If the charger rated output is only 2 to 10 amps, and the battery is deeply discharged, this voltage is normal. The charger is working properly; it just takes a long time to recharge a battery using this charging device.

3. If the charger is rated at 20 amps or higher, watch the battery voltage. If the voltage does not increase to above approximately 13.2 volts, a voltage drop or a problem may have developed in the charging circuit. A diode or isolator in the charging circuit introduces a voltage drop of 0.6 to 1.0 volts. A voltage regulator set at 13.8 volts will only supply 12.8 to 13.2 volts to the battery because of the diode. It takes a long time to recharge the battery with the diode in the circuit. If there is no diode, check all fuses, switches, and connectors for a voltage drop. See "Find a Voltage Drop" on page 162. A component with a voltage drop greater that 0.5 volts should be cleaned or replaced.

4. Using an ammeter, check the amperage output of the charging device. An ammeter must be rated for the maximum amount of amperage supplied by the charging device. Amperage greater than the meter's rated amperage can destroy the ammeter.

Remember, amp-hours are replaced by amperage over a period of time.

The greater the amperage supplied by the charging device, the quicker the amp-hours will be replaced. The amperage, however, must be reduced as the battery nears full charge.

If the amperage supplied by the charger is low and the voltage is high, 13.8 to 14.2 volts, a battery may be nearing full charge.

When charging a deeply discharged battery if the voltage increases quickly and the amperage decreases, the charger is a constant voltage or taper charger. The charger is working properly but reduces the amperage long before the battery nears full charge; it takes a long time to recharge the battery.

TROUBLESHOOTING A BATTERY

Like humans, batteries have a limited life span, with some living longer than others. Batteries will have a long useful life if maintained and charged properly, but if abused, they will die before their time. Some suffer from sulfation and die a slow death. Others develop internal shorts and die quickly. Some just wear out.

Inspect the Battery Terminals, Container, and Electrolyte

1. Check for corroded terminals.

   - Corrosion between the battery terminal and the cable causes resistance, and restricts the current to and from the battery.

   - During charging, the battery will not receive all the voltage supplied by the charging device and will become undercharged.

   - During discharge, the circuit loads will not receive all the voltage supplied by the battery. Keep the terminals corrosion free.

- Internal battery damage may have occurred if the terminals have been hammered, twisted, or abused.

2. Check the container for cracks, abrasions, and cleanliness. Batteries with obvious damage should be replaced, because internal damage could have occurred.

   - Check the straps that hold down the battery to insure that the battery is secure. A battery should not vibrate because vibration can shorten the battery life.

   - Check for foreign material on the battery cover. Dirt, liquids, or scraps of material mixed with acid around and between the terminals can increase the battery self-discharge rate. Impurities can fall into the electrolyte and contaminate it, causing poor performance and a high self-discharge rate. Keep the cover clean.

3. Check the electrolyte.

   - The electrolyte level should be above the tops of the separators or to the fill level of the battery. If not, fill the cells with distilled water and charge for 15 minutes to mix the water with the electrolyte before taking any measurements.

   - Low electrolyte level causes the exposed plates to sulfate. The sulfated areas become permanently inactive, reducing the battery capacity.

   - Low electrolyte level can indicate poor maintenance, overcharging, or internal shorts. Excessive water loss can be a sign of overcharging caused by a high voltage regulator setting or a worn out battery. An internal short in one cell can cause excessive water loss in that cell.

   - Keep all cells properly filled with electrolyte.

   - Insure that the electrolyte is not discolored or contaminated with a foreign material. If the electrolyte is cloudy or muddy, shedding of the active material may have occurred due to overcharging or vibration. If the electrolyte emits an odor or is discolored, an impurity may have contaminated the electrolyte.

- Any impurity in the electrolyte may cause the battery to fail; if so, the battery needs to be replaced.

Testing a Battery Using a Hydrometer

The battery's state of charge is determined by measuring a battery's specific gravity using a hydrometer; see figure 9-4. Specific gravity is the weight of the electrolyte compared to the weight of water, which has a specific gravity of 1.0. The specific gravity of a battery ranges from 1.120 to 1.300 depending on the battery's state of charge. An indication of the battery health is determined by measuring the specific gravity of each cell to determine any differences between cells.

Electrolyte is drawn from a cell into the hydrometer's glass tube by means of the bulb-type syringe. Inside the glass tube is a sealed glass float with a calibrated scale on its stem. When the hydrometer is held vertically, the glass float sinks to a certain level in the electrolyte depending on the density of the acid. Read the specific gravity of the electrolyte where the surface of the liquid comes into contact with the scale—disregard the curvature of liquid against the glass parts. Keep the hydrometer clean to insure that contamination is not introduced into the cells.

When drawing the electrolyte into the hydrometer, insure that enough liquid is in the tube to allow the float to move freely. If there is not enough liquid, the float will give an improperly high reading, indicating a good battery when it may not be. Some hydrometers have an insert to prevent the float from going up into the bulb. If there is too much liquid, the top of the float will hit this insert and hold the float down, indicating a deeply discharged battery when it may not be.

Figure 9-4 Hydrometer

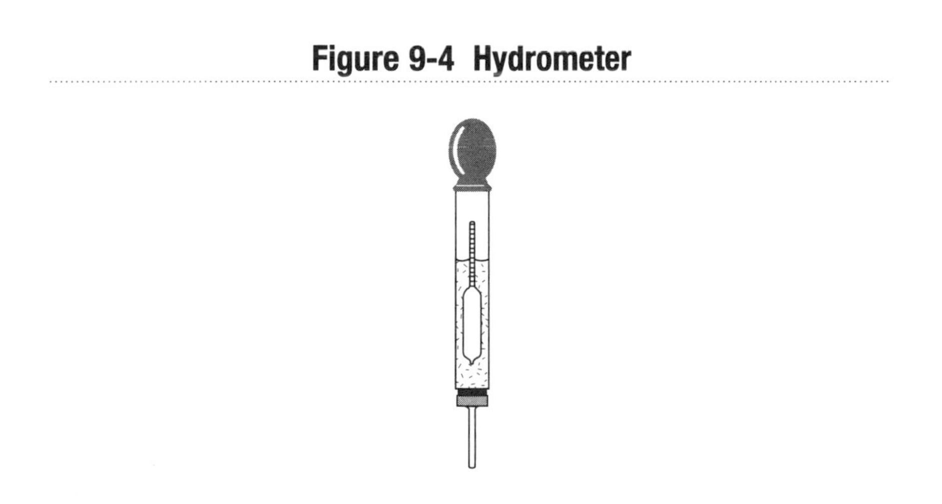

A deeply discharged battery takes a long time to recharge, and it takes several hours before any improvement is noticed in the specific gravity readings. This is because the electrolyte does not mix well until the battery approaches full charge and the battery starts to gas. This gassing mixes the electrolyte, and the specific gravity reading starts to improve.

Temperature effects the readings because the electrolyte expands and becomes less dense when heated, with the bulb sinking lower in the electrolyte. With cooler electrolyte, the opposite happens. A correction must be applied when the electrolyte temperature is not 80°F (26.7°C). Some hydrometers use a reference temperature of 60°F (15.5°C), so check the hydrometer instructions. A correction factor of 0.004 specific gravity or 4 "points of gravity" is used for each 10°F (5.5°C) change in temperature. Add 4 points of gravity to the indicated reading for each 10°F above 80°F and subtract 4 points for each 10° F below 80°F. This is especially important in extreme temperatures because the correction is substantial. A hydrometer with a built in thermometer is the best one to use when reading the specific gravity of batteries.

In temperate climates, the specific gravity reading for most fully charged batteries is in the 1.250 to 1.280 range. In tropical climates, where water never freezes, the specific gravity reading for a fully charged battery is in a range of 1.210 to 1.230. Batteries in extremely cold weather use stronger electrolyte and have a specific gravity for a fully charged battery in the range of 1.290 to 1.300. A battery's service life is shortened by higher concentrations of electrolyte.

Table 9-1 indicates the specific gravity readings that correspond to a battery's state of charge at 80°F.

Table 9-1 Approximate State of Charge

Charged	1.225 Initial Full Charge	1.265 Initial Full Charge	1.280 Initial Full Charge	1.300 Initial Full Charge
100%	1.225	1.265	1.280	1.300
75%	1.185	1.225	1.240	1.255
50%	1.150	1.190	1.200	1.215
25%	1.115	1.155	1.170	1.180
Discharged	1.080	1.120	1.140	1.160

Some types of hydrometers indicate the specific gravity reading without using the decimal point and indicate, for example, 1265 as a battery at 100 percent of capacity.

Hydrometer Test

1. Measure and record the specific gravity of each cell.

2. If the range of readings between the highest and lowest is greater than 50 points or 0.050 specific gravity or the lowest is less than 1.225, charge the battery.

3. Charge the battery until all the cells are greater than 1.225 specific gravity and the range is less then 50 points. If this is not achieved, the battery probably has a failing or defective cell; replace the battery.

Open Circuit Voltage

The battery open circuit voltage is the voltage when the battery is not delivering or receiving power. Allow the battery to rest or stabilize at least 15 minutes—preferably longer—before the open circuit voltage is measured. If the battery has been on charge, place a heavy load on the battery for 30 seconds to remove the "surface charge." This stabilizes the voltage. Other than measuring the battery specific gravity, the open circuit voltage is a good way to determine a battery's state of charge. Use the following table to determine a battery's state of charge when the temperature is between 60°F and 100°F.

Open Circuit Voltage	Percent Charge
12.6 or greater	100%
12.45 - 12.6	75% -100%
12.24 - 12.45	50 - 75%
12.06 - 12.24	25 - 50%
11.7 - 12.06	0 - 25%
11.7 or less	0%

Charge the battery if the state of charge is less than 75%. Unlike measuring the specific gravity, individual cell voltages can not be measured to determine if one cell is weaker than the others. Two 6 volt batteries connected in series to form a 12 volt battery bank can be disconnected and allowed to stabilize before measuring their voltage. If the voltage reading between the two batteries is 0.05 volts or more, the lower voltage battery is weak or failing.

Test for a Battery that Fails to Keep a Charge

1. Fully charge the battery, and record its open circuit voltage and specific gravity.

2. Place it on open circuit for three days by disconnecting the two battery cables.

3. Measure the battery open circuit voltage and specific gravity. An internal short may exist in a cell if the battery open circuit voltage drops by 0.20 volts or its specific gravity decreases by 35 points (0.035) in one or more cells, during the 3 days. If so, replace the battery.

Load Tester

A battery can have high voltage and specific gravity readings but may still not be able to supply a high amperage load. A battery load tester is a specialized device that connects across the battery terminals and creates an extremely high load while measuring the battery voltage. A test load equivalent to 50 percent of the Cold Cranking Amp or 3 times the 20 amp-hour capacity of the battery is placed on the battery for 15 seconds. A battery in good condition will maintain a good voltage reading, while a weak battery voltage drops below 9.6 volts after 15 seconds.

Follow the manufacturer's instructions when doing a load test on a battery. The following is a guideline.

1. Disconnect the battery cables starting with the ground.

2. Measure the electrolyte temperature using a hydrometer that has a built-in thermometer because temperature affects the minimum voltage a battery will drop to during the load test.

3. Set the load tester selector switch to the type of battery or select a range for 50 percent of the CCA of the battery.

4. Connect the load tester leads to corrosion free terminals.

5. Apply the load for 15 seconds. If the voltage drops below the minimum listed voltage on the chart, go to step 7.

Electrolyte Temperature	Minimum required voltage
70°F (21°C)	9.6
60°F (16°C)	9.5
40°F (4°C)	9.3
20°F (-7°C)	8.9

6. If the battery passes the load test, return the battery to service.

7. If it fails, measure the battery's state of charge. Using a hydrometer, measure the specific gravity. Charge the battery if the specific gravity readings are less than 1.225 and the range is less than 50 points. Using a voltmeter, measure the battery open circuit voltage after letting the battery stabilize at least 15 minutes. If it is less than 75 percent of state of charge, charge the battery until its state of charge is greater than 75 percent. After charging the battery, test the battery again using the load tester. If the specific gravity or the state of charge is greater than 75 percent and the battery fails the load test, the battery should be replaced.

Undercharged Battery

1. If a battery never seems to be fully charged, check the battery to see if it is dead or dying. See the section on troubleshooting a battery on page 167.

2. Check for a loose alternator belt. The belt should not depress more than one half inch at the midpoint. If the belt is loose, tighten the belt. Replace if necessary. Also, the alternator needs to be turning at a high speed to generate the proper current.

3. Using a multimeter set at DC volts, check for any voltage drop between the alternator positive post and the battery. See "Find a Voltage Drop" on page 162. If there is a voltage drop, excessive resistance in the charging circuit may prevent proper charging of the battery. Clean and tighten any connectors or terminals that introduce excessive voltage drops. Insure that the ground connections are clean and tight.

4. The voltage regulator may be faulty or set too low. A battery at or near full charge should have a voltage of approximately 14 volts. If not and the circuit components are okay, have a qualified technician check the alternator and voltage regulator.

Overcharged Battery

If a battery continues to require topping off with water, overheats, or gives off an acid smell, it is being overcharged.

1. Check to see if the battery is dead or dying. See the section on troubleshooting a battery on page 167.

2. Check the battery voltage while it is being charged. The voltage of a fully charged battery should not be greater than 14.8 volts. If so, the voltage regulator may be faulty. Have a qualified technician check the voltage regulator.

Determine the Cause of Battery Failure

You need to understand why a battery fails because just replacing the battery may not solve an underlying problem.

1. The age of the battery.

    - A battery may have become worn out because of age. A record of when the battery was sold or put in service is helpful in determining if the battery failed prematurely or simply worn out. The month and year of purchase are usually indicated on the label, or your records should indicate how long the battery has been in service. All battery manufacturers date their products by stamping a date code on the cover or container. The code is usually a string of letters and numbers, but you only need to be concerned with the first two characters. The numbers indicate the year—1 for 1991, 9 for 1989—and the letters in the code indicate the month—A for January, B for February, and so on. Some codes skip the letter I so M indicates December. As an example, E4 indicates May 1994.

2. Battery application.

    - Check to insure that the battery design is correct for the application. An automotive starting battery used as a house battery will fail prematurely.

- Check to insure that the battery is properly sized for the application. If the daily amp-hour requirement is about equal to the battery capacity, the battery will fail prematurely. The amp-hour capacity should be three times the daily amp-hour requirement.

- Check to determine that excessive electrical loads were not added to the vehicle without the battery capacity being increased.

3. Operation of the Battery.

- Check to determine if the battery has been left in an unused vehicle and has not been charged regularly. A vehicle in use regularly cycles a battery, keeping it charged. A battery left for a lengthy period self-discharges and becomes undercharged, a condition that is harmful to the battery life.

- Check to insure that the charging system is operating correctly. An inadequate or faulty charging system will cause a battery to fail prematurely. High voltage can cause overcharging, excessive gassing, and water loss. Low voltage can cause undercharging, with the sulfate becoming hard and difficult to remove. In both cases, the battery capacity will be reduced.

- Check to insure that a deep-cycle house battery is not regularly cycled to below 20 percent of its capacity. Discharging a battery regularly below 20 percent state of charge is harmful to even a quality deep-cycle battery.

ELECTRICAL LEAKAGE

A small electrical leak may cause a battery to be continually discharged when attached to an unused vehicle. A electrical leak is unwanted current flowing to ground, but the current is not great enough to cause a short circuit. The cause of the leak could be corrosion, dampness, poor connections or equipment insulation.

Test for electrical Leakage:

1. Turn off all the circuits in the vehicle.

2. Remove the positive cable from the battery terminal.

3. Using a multimeter set to DC voltage, measure the voltage between the positive battery terminal and the positive battery cable. If no leakage is present, the voltage will be zero (perhaps your multimeter will measure a few thousands of a volt); see figure 9-5. If the voltage is 12 volts, a circuit is still on in the vehicle. Double check all circuits to insure everything is off. Go to step 4 if the voltage remains at 12 volts.

4. Switch the multimeter to measure current and set the DC amp range to the highest setting. The current passes through the multimeter, so if a circuit is left on and it draws more amperage than the multimeter is rated for, the meter could be damaged. Measure the amperage between the positive battery terminal and positive cable. Any reading of 1 amp to 0.01 amp is a major electrical leak. More than about 1 amp indicates that a circuit is still on. A reading of less than this is a minor leak.

5. If an electrical leak is present, clean up the battery terminals, cables, switches, and connections. Pull fuses in the circuits to determine where the leak is occurring. When the leak disappears after pulling a fuse, the leak is in that circuit. Find and correct the leak. Digital clocks are constantly on, but only draw a few hundredths of an amp.

Figure 9-5 Checking for an Electrical Leak with a Voltmeter

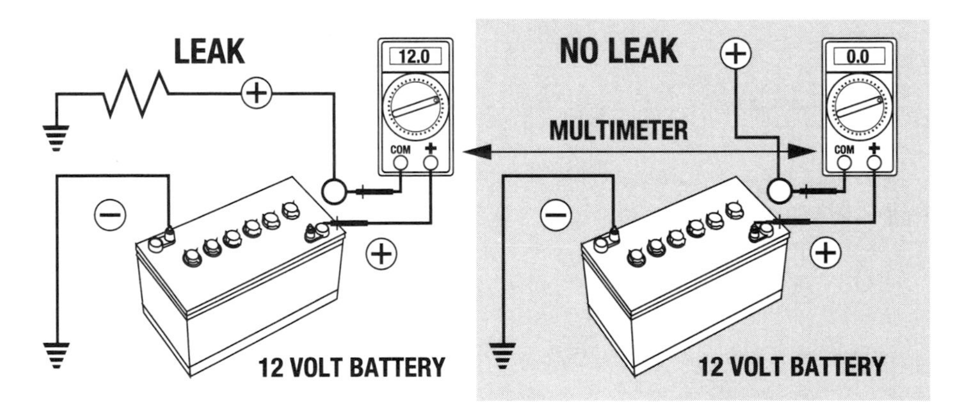

Use an Ohm Meter to Find an Electrical Leak

1. Turn off all circuits on the vehicle. Disconnect the positive cable from the battery. Set the multimeter to the ohms position. Connect the multimeter to the disconnected positive cable and to the battery negative terminal.

2. A resistance measurement of 0 to 10 ohms indicates something is still turned on. Ten to 100 ohms indicates a digital clock or instrument light is on. One hundred to 1,000 ohms indicates a significant leak. A resistance measurement of greater than 10,000 ohms or an infinite reading indicates little or no leak.

3. If an electrical leak is present, clean up the battery terminals, cables, switches, and connections. Pull fuses in the circuits to determine where the leak is occurring. When the leak disappears after pulling a fuse, the leak is in that circuit. Find and correct the leak.

TROUBLESHOOTING AN ALTERNATOR

The alternator is the most common charging device found on RVs and boats. Most alternator outputs are controlled by constant voltage regulators set at between 13.8 and 14.4 volts. Alternators produce from 35 to 200 amps depending on the model, but since they are controlled by a constant voltage regulator, the amperage decreases as the voltage increases. The battery voltage increases to around 14 volts, but even for a deeply discharged battery the amperage will drop below the alternator's rated output. Multi-stage voltage regulators follow the charging sequence discussed in Chapter 5.

This troubleshooting guide is to determine if a problem exists with the charging circuit or with the alternator. If it is determined that the alternator or voltage regulator is bad, it is best to have a qualified technician check it out.

On most RVs and boats, the alternator charging circuit charges the starting battery and the house battery. An isolator with a diode, a solenoid, or a battery selector switch is usually mounted in the circuit to isolate the batteries, so the starting battery is not discharged along with the house battery. If the alternator is charging the starting battery but not the house battery, the problem is in the house battery charging circuit and not with the alternator.

An Alternator with No Output

1. Start the engine. Measure the starting battery voltage at the battery terminals using a multimeter set on DC volts. The voltage should increase to around 14 volts. If yes, go to step 3; the alternator is working.

2. If the starting battery voltage remains around 12 volts after the engine is started and run for a few minutes, a problem exists with the alternator or the charging circuit.

   A. Turn off the engine. Inspect all wiring and connections. Insure that the battery terminals and ground point are tight and clean.

   B. Check the alternator drive belt to insure it is not loose. The belt should not depress more than one half inch at the midpoint. If the belt is loose, tighten the belt. Replace if necessary.

   C. Insure that no other charging device, solar panel, or battery charger, is charging the battery.

   D. If a battery selector switch is in the charging circuit, insure that it is on the correct setting.

   E. Turn on the ignition switch, but do not start the engine.

   F. Check the battery voltage. It should be approximately 12 to 13 volts.

   G. Check the voltage between the alternator positive terminal and ground; see figure 9-6. If the circuit is good, the voltage at the alternator should be the same as the battery, or if an isolator is in the circuit the alternator voltage will be zero. If not, a problem exists in the circuit between the alternator and the battery. Check for an open in the charging circuit.

   H. Start the engine. Check the voltage between the alternator output and ground. It should be 13 to 14 volts. If not, have the alternator checked by a qualified technician.

Figure 9-6 Checking the Voltage at an Alternator

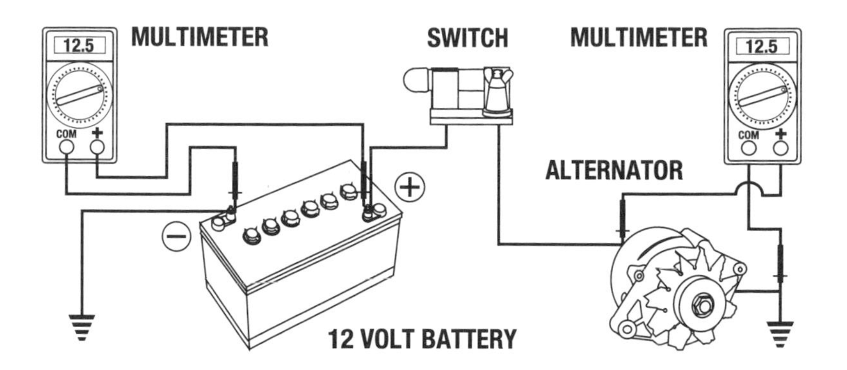

3. Check the voltage of the house battery. The house battery should have a voltage between 13 and 14 volts when the engine is running. If not, there is a problem in the house battery charging circuit.

   A. If an isolator is in the house battery charging circuit, the house battery voltage will be about 0.6 volts less than that of the starting battery. If the house battery voltage does not increase when the engine is started and remains about 12 volts, troubleshoot the isolator. See the section on troubleshooting isolators on page 180.

   B. If a solenoid is in the circuit, figure 9-7, the house battery voltage will be about 0.2 volts less than that of the starting battery. If so, the solenoid is working properly. If the house battery voltage does not increase when the engine is started and remains about 12 volts, troubleshoot the solenoid. See the section on troubleshooting a solenoid on page 183.

   C. If the isolator or solenoid is okay, check all connections, wiring, and switches for an open circuit. See "Find a Break in an Open Circuit" on page 156.

   D. The wiring and connections in the charging circuit can cause a voltage drop. The longer the wiring runs, and greater the voltage drop will be. If the isolator or solenoid checks out okay but the house battery voltage is more than 1 volt less than the starting battery voltage, check for a voltage drop. See "Find a Voltage Drop" on page 162.

E. Check the house battery to see if it is defective. See the section on troubleshooting a battery on page 167.

Figure 9-7 Check Voltage of House and Starting Battery

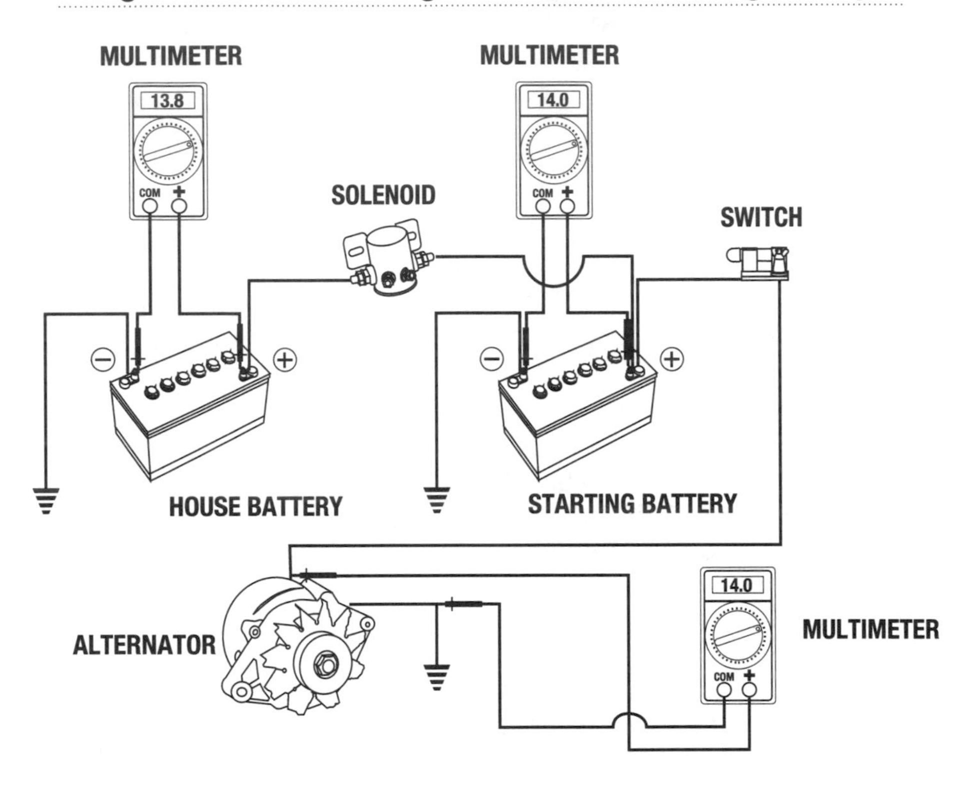

TROUBLESHOOTING ISOLATORS AND DIODES

Isolators consist of diodes that allow electricity to flow in one direction but not in the other. They are mounted in a heat sink that dissipates the heat built up by the diodes. Isolators are used to isolate electrically the starting battery from the house batteries and to isolate several house batteries, so a fully charged battery can not discharge into a drained battery. Isolators have three or four terminals: one where the wiring from the alternator is attached and the other two or three terminals where the cables for the batteries are attached.

Diodes have two terminals: one connected to the charging source, and the other leading to the battery.

Several methods are used to test an isolator.

Testing Isolators and Diodes

1. Turn the engine off.

2. Using a multimeter set on DC voltage, place the red probe on the isolator terminal leading to the alternator and the black probe to ground. The voltage should read from zero to 10 volts. If the voltage is 12 volts, check to insure a battery or a charging device is not attached to this terminal if not, the diode is shorted.

3. Place the red probe on one of the isolator terminals leading to a battery, and the black probe to ground. The voltage should be the same as the battery voltage. Check both battery terminals on the isolator. The reading at the alternator terminal should not have the same voltage as the voltage at the battery terminal or the isolator is shorted.

4. Start the engine.

5. Measure the battery voltage by placing the red probe on the isolator terminal for the battery and the black probe to ground; see figure 9-8. If another battery is attached to another isolator terminal, check its voltage, also. The battery voltages should increase to above 13 volts and be approximately the same. Now measure the voltage at the alternator terminal by placing the red probe on the isolator terminal leading to the alternator and the black to ground. It should read 0.6 to 1.0 volts higher than the battery voltages. The differences in voltage between the alternator and the batteries are the voltage drops caused by the diodes.

If the battery voltage does not increase after the engine is started, the isolator has an open and needs to be replaced.

The battery voltages should read between 13.8 and 14.2 volts. With an isolator in the circuit, the alternator output should be about 15 volts and not between 13.8 and 14.2. If the batteries' voltage is between 13.2 and 13.6 volts, the voltage regulator is sensing the alternator output and not the battery voltage. The batteries are not being properly charged. Talk to a qualified technician to determine if it is possible to change the sensing wire of your voltage regulator to sense the battery voltage and not the alternator output. It may be possible to place a diode in the voltage regulator's sensing wire that would simulate the isolator voltage drop. Then the alternator would produce about 15 volts and the batteries would be charged with the proper voltage even though an isolator is in the charging circuit.

Figure 9-8 Checking the Voltage at an Isolator

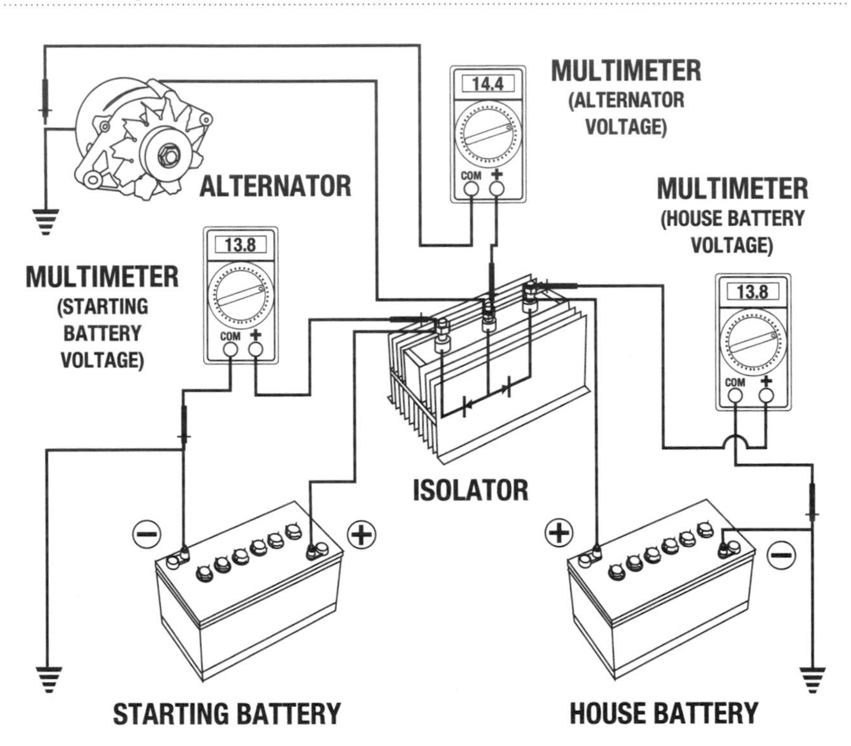

Testing Isolators or Diodes with an Ohm Meter

1. Remove all wires from the isolator or diode. The multimeter's internal battery supplies the current for this test and the voltage from a battery or charging device can damage the multimeter.

2. Set the multimeter on the ohms scale. Hold the red probe to the alternator terminal of the isolator, and the black probe on a battery terminal. A good isolator indicates a low resistance reading between the two terminals. (On some multimeters, the probes are reversed for this test.)

3. Reverse the probes. A good isolator indicates no current (an infinite or extremely high resistance reading on the multimeter). If the ohm meter indicates current (low resistance reading), the diode is shorted. A good diode indicates current in one direction but not the other. If current does not flow in either direction, an open circuit exists in the diode. In both situations, the diode needs to be replaced.

TROUBLESHOOTING A SOLENOID

On some RVs, a solenoid is used to isolate the starting battery from the house battery. A solenoid is an electrical switch that is activated when the ignition key is turned on. This allows the alternator to charge the house battery. When the ignition key is turned off, the starting battery is isolated from the house circuits, preventing the starting battery from being accidentally discharged.

Testing a Solenoid

1. Turn the ignition key on, but don't start the engine. A click should be heard from the solenoid. (Some solenoids, however, may not operate until the engine is running.) If no click is heard, measure the voltage at the solenoid's small terminals. On a solenoid with only one small terminal, the voltage is measured from this terminal to ground. If no voltage is measured, an open circuit condition exists in the ignition circuit. (Some solenoids, used to disconnect the batteries from the distribution panel or load, indicate 12 volts only when the isolation switch is activated, so the 12 volts is not on continuously. On these solenoids, 12 volts will not be indicated when measuring between the two small terminals even when the solenoid is working, but you should hear a click when the solenoid is activated.)

2. If a click is heard or 12 volts is measured at the small terminals, start the engine.

3. Measure the house battery voltage. The house battery voltage should increase to above 13 volts if the solenoid is good. If the house battery voltage does not increase, perhaps the solenoid is bad.

4. Stop the engine and remove the wires from the two large terminals of the solenoid; see figure 9-9.

5. Turn the ignition key on, but don't start the engine. A click should be heard from the solenoid. Using a multimeter set on the ohms scale, place one probe on each large terminal. An infinite amount of resistance indicates the solenoid is not closing and needs to be replaced. If the reading is zero, the solenoid is good.

6. Turn the ignition key off. An infinite amount of resistance indicates the solenoid is opening and isolating the batteries.

7. Continuity can also be checked. Continuity indicates the solenoid is working; no continuity indicates the solenoid is bad.

Figure 9-9 Checking the Continuity of a Solenoid

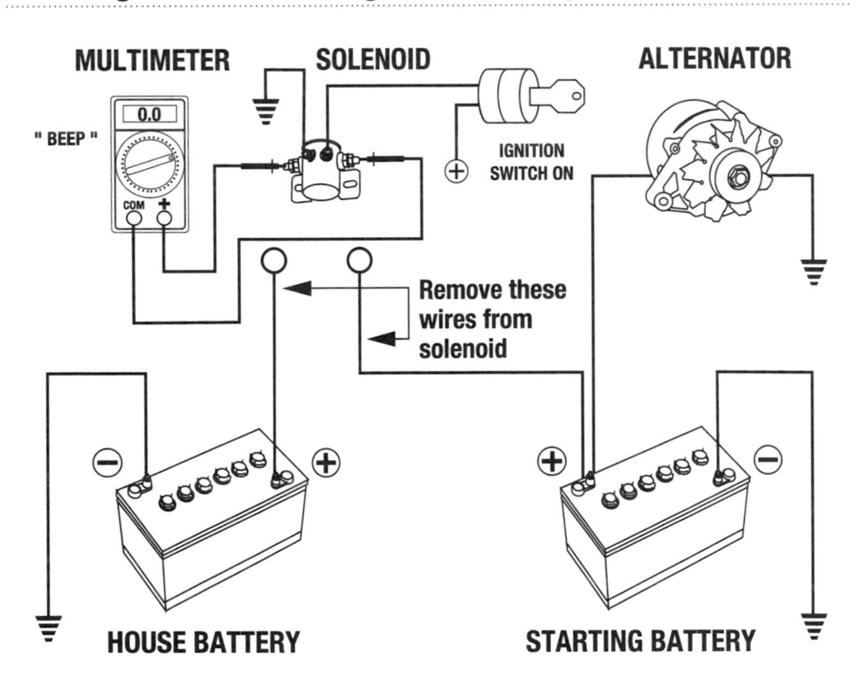

TROUBLESHOOTING BATTERY CHARGERS

Battery chargers (converters) transform 120 volts AC to 12 volts DC to power the 12 volt loads on your vehicle or to charge your batteries. Battery chargers come in various levels of sophistication from simple unregulated trickle chargers to high output multi-stage chargers. All battery chargers are powered by 120 volts AC that can **Kill.** Be careful when working with them. The voltage output from the battery charger is measured at the battery terminals and should be 13.2 to 14.4 volts.

Battery Charger with No Output

1. Check to insure that 120 volts AC is reaching the charger.

    A. Using a multimeter set on AC voltage, check the AC supply you are using either at the AC outlet at a campground or marina, the AC output of a portable generator, or at an AC outlet on the vehicle if an onboard generator is being used. The voltage should be between 110 to 125 volts. If lower, the battery charger may not compensate for the variation of incoming voltage, and its DC output will also be low.

    B. Check all fuses, circuit breakers, switches, connections, and wiring. A short may be internal to the charger. If the battery charger's fuse or circuit breaker blows or trips immediately when plugged in, have a qualified technician check the unit.

    C. AC connections terminate inside the battery charger.
    Caution: Unless you fully understand the circuitry of the charger, do not take off the battery charger cover while attached to 120 volts AC. Go to the next step if the above tests indicate that the AC power is okay. If not, determine why the AC power supply is not working properly.

2. Check the battery leads to the battery to insure that they are clean and tight and have no breaks.

3. Check the battery. If the battery is fully charged, the charger may not come on.

    - Check the battery open circuit voltage. If it is above 12.6 volts, the battery is fully charged. Connect a load of 10 amps or more to discharge the battery. The battery charger should cycle on, and the battery voltage should increase.

- If the battery is deeply discharged or defective, the charger may not come on. Some battery chargers will not cycle on unless the battery has a minimum voltage. Check the battery as described in the section on troubleshooting batteries on page 167. Attach a known good battery to the charger. If the battery charger now cycles on, the problem is not the battery charger but the battery.

4. Check the charger's instruction manual for any internal controls. Battery chargers have various internal control circuits that turn off the charger to protect the battery or the charger. The following are examples of internal controls found on some battery chargers.

   - A **thermal protection switch** shuts down the charger if it is too hot. Turn off the charger, or reduce the DC load, allowing the charger to cool before trying the charger again.

   - An **ignition protection switch** shuts down the battery charger when the engine is started and the alternator is charging the batteries.

   - A **high amperage or overload shutdown switch** turns off the battery charger if the battery is drawing more amps that the battery charger can produce. Reduce the DC loads, check for a defective battery, or check the vehicle's DC equipment for shorts.

   - **Automatic battery chargers** cycle off when the batteries are at a set voltage and cycle on when the battery voltage drops to a certain level. One battery in a multi-battery charging system may be the control battery. The charger may not cycle on unless the control battery voltage decreases, even if the other batteries are deeply discharged. Discharge the control battery until the charger cycles on.

TROUBLESHOOTING AN RV CONVERTER/BATTERY CHARGER

The converter/battery charger on most RVs does not need a battery in the circuit to power most 12 volt lights and motors. The converter transforms 120 volts AC to 12 volts DC and supplies this power to 12 volt lights and motors, bypassing the battery. The DC distribution panel, also part of this unit, takes the 12 volts from the converter directly to fuses that lead to the lights and motors on the RV. Certain 12 volt equipment—TVs, radios, stereos, unfiltered fluorescent lights—can only operate on battery power so their power must come directly from the battery. Depending on the model, converters are designed to supply from 25 to 50 amps to the RV's 12 volt lights and motors.

An automatic relay instantly switches all the 12 volt loads of the RV to the battery when the 120 volt AC power is removed. Once the 120 volt AC is restored, the relay switches the power for the lights and motors from the battery to the converter. A loud click can be heard each time the automatic relay switches.

A battery charger is an option on some models. When the 120 volt AC is connected and the battery is below full charge, the battery charger charges the battery. The battery charger output is stated at 3 amps increasing to 6 amps if the battery is deeply discharged.

Troubleshooting an RV's Converter/Battery Charger

1. Check to insure that 120 volts AC is reaching the charger; see figure 9-10. Check the AC supply using a multimeter set on AC voltage either at the AC outlet at the campground, or at an AC outlet on the RV if an onboard generator is being used. The voltage should be between 110 to 125 volts. If lower, the battery charger may not compensate for the variation of incoming voltage and its DC output may be low.

2. Check the AC circuit breaker, located in this unit, to determine if it has tripped. If the circuit breaker is tripped, reset the circuit breaker. If it immediately trips, a short circuit condition may exist.

3. Turn off the 120 volt AC. A loud click should occur indicating that the relay is switching the 12 volt loads to the battery. The relay may be bad if a click in not heard.

Figure 9-10 RV Converter/Battery Charger

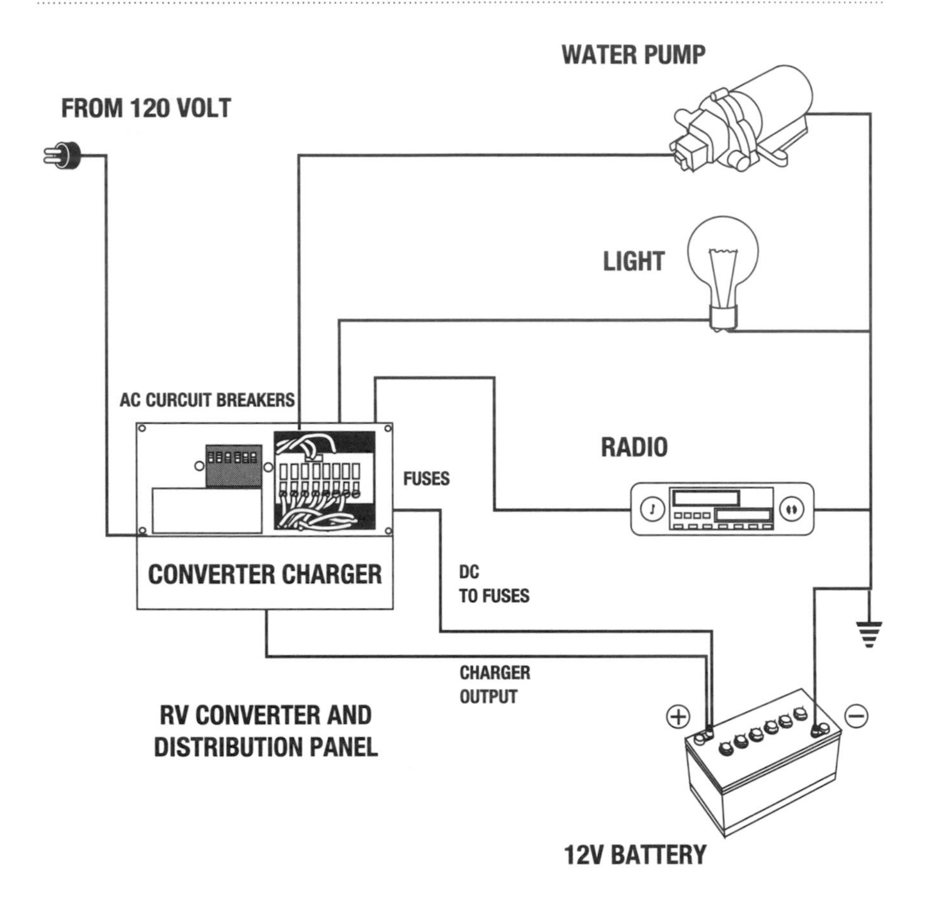

4. Some units have a protective thermal breaker that will shut off 120 volt AC power to the converter if it becomes overheated. The relay switches the load to the battery if this occurs. The reason the converter overheats may be caused by too many 12 volts devices on or an obstruction blocking air to the unit. If overheating is caused by an overload of 12 volt devices, and the relay switches the load to the battery, the battery will be completely discharged if the larger amperage loads are not turned off. The thermal breaker resets itself after the unit has cooled.

5. If the problem is that the battery is not supporting the 12 volt loads, the reason is very likely that the battery is not being adequately charged. The battery charger output is only 3 to 6 amps. Check the battery voltage when 120 volts AC is supplied to the unit. The battery charger is working if the battery voltage increases to above 13 volts, but at 3 to 6 amps it's going to take a very

long time to recharge the battery. If the battery voltage does not increase, a problem exists with the charger. Does the relay click when the 120 volt AC is turned on? If not, the relay is bad. Have a qualified technician check the unit. Some models of this converter/battery charger do not have the optional battery charger, so the battery is not charged by this unit. Some other method must be used to recharge the battery.

6. If the lights and the water pump work but the radio does not when plugged into 120 volts AC, the problem may be a dead battery . The radio is not powered directly by the converter. The radio requires 12 volt battery power, whereas the lights are powered directly by the converter.

7. Check the battery. See the section on troubleshooting batteries on page 167.

TROUBLESHOOTING SOLAR PANELS

If a battery is not being charged by a solar panel, check the following:

1. Check the solar panel's case for damage. Is the case cracked or broken, allowing moisture into the solar cells? Are there signs of corrosion on the solar cells? If the case is plastic, is there ultraviolet degradation? Corrosion of the solar cells is impossible to fix. Replace the solar panel.

2. Measure the solar panel voltage using a multimeter set on DC volts. Disconnect the leads at the panel, and measure the voltage. The voltage should be 15 volts or greater in bright direct sunlight. If not, the internal connections are defective.

3. Check the fuses, connectors, and wiring for breaks or corrosion. Check for resistance using a multimeter set on ohms; a continuity check can also be made. Cover the solar panel, so it does not produce voltage during a resistance or continuity test. The solar panel voltage is enough to damage the multimeter when in the ohms position.

4. Check the blocking diode (see the troubleshooting section for isolators and diodes on page 180) for an open that prevents that electricity from passing. Replace the diode if necessary.

5. If the solar panel has a voltage regulator, check the voltage of the regulator. If the solar panel is producing voltage but there is no voltage after the regulator, the regulator is defective.

TROUBLESHOOTING WIND GENERATORS

Wind generators consist of a propeller driving an alternator or DC motor, and a circuit with a regulator, a diode, various fuses, switches and wiring. All must be checked if the voltage at the battery is below 13 volts.

1. Check for blown fuses and switches set incorrectly.

2. Check the battery voltage when the generator is operating, using a multimeter set at DC volts. The voltage should be greater than 13 volts. If so, the generator is working.

3. Check the voltage output at the wind generator. The open circuit voltage can be very high, more than 30 volts, and can give a severe shock. If voltage is measured at the generator, it is working. If not, stop the wind generator and short the two wires from the generator together. Turn the propeller by hand. It should be difficult to turn the propeller blades by hand if the generator is okay. If not, the generator is defective.

4. Using a multimeter set to the ohms function, check the resistance of all the wiring, connectors, and switches. The resistance should be zero or a few tens of an ohm. If you measure an infinite amount of resistance, a break is in the circuit. A continuity check can also be made. The wind generator can not be turning during a resistance or continuity test. The voltage output of the wind generator is enough to damage the multimeter.

5. Check the diode (see the troubleshooting section for isolators and diodes on page 180) for an open that prevents electricity from passing. Replace the diode if necessary.

6. Check the voltage output of the regulator. If the wind generator is producing voltage but there is no voltage after the regulator, the regulator is defective.

Glossary

Absorption stage: second stage of a multi-stage charging cycle where the voltage has increased to 14.2 volts and the current has decreased to the amount the battery will naturally accept.

Active Material: the material on the battery plates that reacts with the sulfuric acid in the electrolyte during the charging and discharging cycles of a lead-acid battery. The active material is lead dioxide on the positive plates and sponge lead on the negative plates.

Alternator: an electromechanical device on an engine that transforms mechanical energy into electrical energy by rotating a magnetic field within a circle of stationary copper wire windings. The alternator produces alternating current that is rectified by diodes into direct current, which is then used to recharge the batteries and power the vehicle's electrical loads.

Alternating Current (AC): an electrical current that does not continually flow in one direction like direct current, but reverses its direction at regular intervals; each reversal is a cycle. The number of cycles per second is the frequency of the alternating current. In the United States, the frequency is 60 cycles per second or 60 Hertz; in Great Britain, it is 50 Hertz.

Ammeter: an instrument for measuring electrical current. It must be placed in the current path so the flow is through the meter. Clamp-on ammeters measure the magnetic fields surrounding a circuit.

Ampere (Amp): the unit of measurement for electrical current. Since billions of electrons flow through a circuit, the coulomb represents a group of electrons or 6.25×10^{18} electrons. A current through a circuit of one coulomb per second is one ampere. Ampere has a rate or time element, the second, just like gallon per hour has a time element.

Ampere-Hour (Amp-hour, Ah): A unit of measure for a battery's electrical storage capacity. The amp-hour is determined by multiplying the current in amperes by the time in hours of discharge or charge. If a battery is discharged by 20 amps in one hour, the battery's electrical storage capacity is reduced by 20 amp-hours. If a battery is charged by 5 amps over a 4 hour period, the battery's electrical storage capacity is increased by 20 amp-hours.

Battery: a group of two or more cells, connected in series, that converts chemical energy into electrical energy by electrochemical reaction. Three cells connected in series is a 6 volt battery, and six cells connected in series is a 12 volt battery.

Battery Selector Switch: a switch installed in an electrical circuit between two or more batteries or battery banks, which allows the batteries to be paralleled, only one battery bank selected for use, or disconnected from the electrical loads.

Bulk stage: first stage of a multi-stage charging system during which a deeply discharged battery will accept a large amount of current. The stage ends when the battery starts to gas.

Bus Bar: A metal bar used to make multiple connections.

Capacity: a battery's electrical storage capability as measured in amp-hours. The amp-hour capacity of a fully charged battery is its ability to deliver a specified quantity of electricity (amp-hour, Ah) at a given rate (amps) over a definite period of time (hr). For example, a fully charged 100 amp-hour battery delivers 5 amps for 20 hours. The capacity of a battery depends upon a number of factors: the active material's weight, amount, alloys, and density; the plate design, number, spacing, and dimensions; the electrolyte specific gravity, and quantity; the separator design; the discharge rate; the temperature; the final limiting voltage; the internal and external resistance; and the age and life history of the battery.

Cell: the basic electrochemical current producing unit that consists of a set of lead dioxide or positive plates and a set of sponge lead or negative plates, immersed in a solution of sulfuric acid or electrolyte, separated by separators and placed in a container. Each cell has a voltage of approximately 2 volts.

Circuit: An electrical circuit is a path of electrical current. A closed circuit has a complete path for the electrical current to follow, as in a circle. Like a circle, the electrical path is a continuous or endless loop. An open circuit has a broken or disconnected path.

Circuit Breaker: an electromagnetic switch used to protect the circuit from excessive current. The current breaker opens (trips) the circuit if the ampere rating of the breaker is exceeded.

Cold Cranking Amps (CCA): A battery rating for starting batteries that indicates the number of amps a lead acid battery can deliver for 30 seconds at 0°F (-17.8°C) and maintain, at least, a voltage of 1.2 volts per cell, or 7.2 volts for a 12 volt battery. The higher the Cold Cranking Amps the more power the battery can deliver to start an engine.

Corrosion: A foreign substance such as rust or tarnish deposited by chemical action. The sulfuric acid of the electrolyte is corrosive to metal fittings, such as battery terminals.

Current (I): the rate of flow of electricity or the movement of electrons along an electrical path. The ampere (amp) is the unit of measure for electrical current: one coulomb/second, or 6.25×10^{18} electrons/second.

Deep-cycle Battery: designed to be discharged to 20 percent of its amp-hour capacity because it is constructed with thicker, stronger plates, with additional active material, and with heavy separators. Six volt golf cart batteries are examples of deep-cycle batteries.

Diffusion: the process by which the sulfuric acid in the electrolyte dissipates or flows to the active material on the plates or to the outer reaches of the cell.

Diode: a solid state device that allows current to flow in only one direction: an electronic check valve.

Direct Current (DC): An electrical current that flows in only one direction in an electrical circuit. A battery produces direct current and must be recharged with direct current from the opposite direction.

Discharging: A battery is discharging when it is delivering current; its state of charge is being reduced.

Electricity: a form of energy produced electrostatically, mechanically, chemically, or thermally, resulting in electrons flowing through a circuit.

Electrolyte: a diluted solution of sulfuric acid and water in a lead acid battery. It is a liquid that conducts electricity between the plates, and supplies water and sulfate for the electrochemical reactions that take place.

Electromagnet: an iron core wrapped with a wire coil. A magnetic force is induced by passing a current through the coil.

Electromotive force: another name for voltage.

Electron: an elementary particle with a negative charge.

Equalization charge: the process of driving a wet cell lead acid battery to its highest natural voltage to reconvert all the lead sulfate on the plates to active material and sulfuric acid. Battery cells show differences in their state of charge because of temperature, construction, electrolyte, and degree of sulfate on the plates. An equalization charge "equalizes" the cells to the same state of charge.

Float Charge: the final stage of a multi-stage charging system during which a small amount of current is supplied to the battery keeping it at full charge, without overcharging.

Fuse: a protective device containing a conductor that melts if current goes above the fuse's rating—breaking the electrical circuit.

Gassing: A process where the battery electrolyte breaks down, giving off hydrogen at the negative plates and oxygen at the positive plates.

Grid: a lead alloy frame that the active material of the plates of a battery is attached to and through which the electrical current is conducted to other plates and the battery terminal.

Ground: A path for electrical current. The ground on an RV is the chassis and on a boat it is the engine block or a grounding strap. One battery cable, usually the negative cable, attaches to ground and all negative wires from the loads must also attach to ground to complete the electrical circuit.

Hertz: one cycle per second. AC circuits have 60 Hertz in the United States and 50 Hertz in Great Britain.

Hydrometer: an instrument used to determine the state of charge of a battery by measuring the specific gravity, or concentration of sulfuric acid, in the electrolyte.

Leakage: undesired flow of electricity around or through a circuit's power source, connections, or loads, usually to ground.

Load: A device that consumes power and does the work in an electrical circuit.

Load Tester: an instrument that determines a battery's ability to perform under load by discharging the battery while measuring the battery voltage.

Magnetic Field: The force field around a magnet.

Ohm: measure of electrical resistance.

Ohms Law: expresses the relationship between volts, amperes (current), and resistance (ohms) in an electrical circuit. Ohms law states that one volt (V) can move one ampere of current (I) against one ohm of resistance (R). This relationship is expressed in equations: $V = IR$, $I=V/R$, $R=V/I$. If two values are known, the third can be calculated.

Open circuit: an incomplete or broken current path.

Open circuit Voltage: of a battery is the terminal voltage when the battery is not being discharged—delivering energy, or being charged—receiving energy.

Parallel Circuit: An electrical circuit that provides two or more paths for the electrical current to follow. The electrical circuit on an RV or a boat is an example of a parallel circuit; one battery provides electrical energy for many different electrical loads. Each path or branch of a parallel circuit can contain a load that requires a different amount of electrical current; a battery provides 200 amps to a starter but only one amp to a light bulb.

Potential Difference: The voltage difference between two points.

Reserve Capacity rating: A battery rating developed to represent the time it would take a starting battery to become discharged while operating all the essential loads on a vehicle after the alternator failed. It measures the time in minutes that a battery can deliver 25 amps at 80° F to a voltage end point of 10.5 volts. The greater the reserve capacity rating the longer a 25 amp load will operate.

Resistance: in an electrical circuit is the opposition to the flow of electrons or the current: the greater the electrical resistance, the greater the opposition to the electrical current. Electrical resistance is measured in ohms.

Separator: material placed between a cell's positive and negative plates that separates them but allows the current and electrolyte to pass through.

Series Circuit: an electrical circuit with only one path for the current to follow. A flashlight is an example of a series circuit; the batteries provide electrical energy to only one light bulb. Two 1.5 volt flashlight batteries are placed in series, so they can support a 3 volt light bulb.

Short Circuit: an unintentional current path that bypasses the circuit electrical load. The bypass is usually of very low resistance and causes a large amount of current to pass through the circuit—more current than it is designed to handle. The result is a circuit failure: a fuse blows, a circuit breaker trips, or a component melts. A short in a battery cell may be permanent enough to discharge the battery.

Shunt: a device used in conjunction with an ammeter to measure current. It has a known low-resistance. Current through the shunt causes a voltage drop—the greater the current the greater the voltage drop. The voltage drop is a measure of the amperage through the shunt.

Specific Gravity: the density of a liquid compared with the density of pure water, which has a specific gravity on 1.0. The electrolyte specific gravity is determined by comparing its weight to an equal volume of water. The sulfuric acid in the electrolyte results in a higher specific gravity reading. A high specific gravity reading indicates a high state of charge for the battery.

State of Charge: the amount of electrical energy stored in a battery expressed as a percentage of the capacity when fully charged. A battery at 60 percent of charge means 60 percent of the battery energy is still available and 40 percent has been used.

Volt or voltage (V): the unit of measure of electrical potential difference or pressure.

Voltage Drop: The net difference or the "loss" of voltage measured across a circuit component. Unwanted resistance in wiring, connections, and switches causes voltage drops. Therefore, the voltage is reduced at the electrical load, preventing the load from operating properly.

Watt: the unit for measuring electrical power generated or consumed in an electrical circuit. Watts or power of a circuit is the rate of doing work—moving electrons by an electrical potential or voltage. Watts or power is determined by multiplying circuit voltage times its amperes: Watts (power) = Volts x Amperes.

Watt-Hour (Watt-hr, Wh): the unit of energy equal to one watt operating for one hour.

Bibliography

Angerbauer, George. *Principles of DC and AC Circuits.* North Scituate Mass: Durbury Press, 1978.

Battery Council International. *Battery Service Manual,* 10th ed. Chicago: Battery Council International, 1987.

Berndt, D. *Maintenance for Butteries: A Handbook of Battery Technology.* Somerset, England: Research Studies Press Ltd, 1993.

Brant, Bob. *Build Your Own Electric Vehicle.* Blue Ridge Summit, Pa: Tab Books division of McGraw-Hill, 1994.

Bureau of Naval Personnel, *Basic Electricity,* New York: Dover Publications, 1970.

Calder, Nigel. *Boatowner's Mechanical and Electrical Manual.* Camden, Me: International Marine, 1990.

Evans, Alvis J,. *Using Your Meter: VOM and DVM Multitesters,* 2th ed. Richardson, Texas: Master Publishing Inc. 1994.

Fink, Donald G. ed, *Standard Handbook for Electrical Engineers.* New York: McGraw Hill, 1978.

Harter, James H. and Paul Y. Lin. *Essentials of Electrical Circuits.* Reston Virginia: Reston Publishing Co, 1982.

Heiserman, David L. Revisor. *Understanding Electricity and Electronic Circuits,* New York: Howard W. Sams and Co., A division of Macmillian Inc., 1987.

Hickey, Henry V. and William Villines. *Elements of Electronics.* New York: Gregg Division of McGraw-Hill, 1980.

Jackson, Herbert. *Introduction to Electric Circuits,* 7th ed. Englewood Cliffs, New Jersey: Prentice Hall, 1986.

Jeffery, Kevin and Nan Jeffery. *Boatowner's Energy Planner.* Camden, Me: International Marine Publishing/Seven Seas, 1991.

Linden, David, ed. *Handbook of Batteries and Fuel Cells.* New York: McGraw Hill Company, 1984.

Mantell, C. L. *Batteries and Energy Systems.* New York: McGraw-Hill Book Company, 1970.

Middleton, Robert G. *Practical Electricity,* 4th ed. Revised by L. Donald Meyers. Indianapolis: Theodore Audel and Company, 1983.

Perez, Richard A. *The Complete Battery Book.* Blue Ridge Summit, Pa: Tab Books, 1985.

Smith, G. *Storage Batteries, Including Operation, Charging, Maintenance, and Repair,* 2th ed. London: Pitman Publishing, 1971.

Smead, David and Ruth Ishihara. *Living on 12 Volts with Ample Power.* Seattle: Rides Publishing Company, 1992.

Vinal, George Wood. Storage Batteries: *A General Treatise on the Physics and Chemistry of Secondary Batteries and their Engineering Applications,* 4th ed. New York: John Wiley and Sons, 1955.

Zeveke, G. and P. Ionkin, A. Netushel, S. Strakhov. *Analysis of Electric Circuits.* Translated from the Russian by Boris Kuznetsov. Moscow: Mir Publishers, 1969.

List of Manufacturers

Your local dealer may not carry or know about some of the electrical products discussed, so the following is a partial list of equipment manufacturers.

Heart Interface Corp.
21440 68th Avenue South
Kent WA 98032
206-872-7225
Fax 206-872-3412
Inverters with multi-stage battery chargers

Trace Engineering
5916 195th N.E.
Arlington WA 98223
360-403-9513
Fax 360-435-2229
Inverters with multi-stage battery chargers

Statpower
7725 Lougheed Hwy.
Burnaby BC Canada V5A 4VB
604-420-1585
Fax 604-420-1591
Multi-stage battery chargers

Professional Mariner
2970 Seaborg Ave
Ventura CA 93003
805-644-1886
Fax 805-644-1895
Multi-stage battery chargers

Ample Power
1150 NW 52nd Street
Seattle, WA 98107
800-541-7789
Fax 206-789-9003
Alternators with multi-stage Regulators
Amp-hour meters

Cruising Equipment Co
6315 Seaview Ave N.W.
Seattle, WA 98107
206-782-8100
Fax 206-782-4336
Amp-hour meters

Balmar
27010 12th Ave NW
Stanwood, WA 98292
360-629-6100
Fax 360-629-3210
Alternators with multi-stage Regulators
Amp-hour meters

Everfair Enterprises
2520 N.W.16th Lane #5
Pompano Beach, FL 33064
954-968-7358
Wind Generators
Solar Panels
Battery Chargers
Inverters

Jack Rabbit Energy Systems
425 Fairfield Ave
Stamford, CT 06902
203-961-8133
Fax 203-358-9250
Wind Generators
Alternators
Monitors

Windstream Power Systems
One Mill Street/P.O. Box 1604
Burlington, VT 05402-1604
802-658-0075
Fax 802-658-1098
Wind generators

Solarex
630 Solarex CT
Frederick, MD 21701
301-698-4200
Solar Panels

Siemens Solar
Box 6032
Camarillo, CA 93010
805-482-6800
Solar Panels

Solec International, Inc
12533 Chadron Ave
Hawthorne, Ca 90250
310-970-0065
Solar panels

RV Power Products
1058 Monterey Vista Way
Encinitas, CA 92024
800-493-7877
RV multi-stage battery chargers
Inverters with multi-stage chargers

Index

A

Absorption rate, 36, 39, 46
Absorption stage
  definition, 191
  description, 64
  solar panels, 72
AC usage, estimate, 12
Active material, 22, 191
Alternating current (AC), 191
Alternator, 58
  charging, 102
  definition, 191
  description, 58
  multi-stage regulator, 66, 112, 117
  selector switch, 87
  troubleshooting, 177-180
Ammeter, *see also* Multimeter, Monitor
  application, 79, 98
  current measurement, 142
  definition, 191
  description, 78
  electrical panel, 81
  installation, 79
  shunt, 80
Amp-hour meter, 81
Amperage, *see* current
Ampere-hour, 6, 191
Amperes, 6, 191
Antimony
  amount in batteries, 45
  description, 45
  self-discharge effect, 30
Appliances
  AC list, 13
  DC list, 9
  estimation of usage, 9, 12
AWG
  (American Wire Gauge), 83
  wire size table, 83

B

Battery
  absorption rate, 37, 39, 46
  active material, 22, 191
  age, 30, 174
  antimony, *see* antimony
  calcium, 44

Battery *(continued)*
- cell, 23, 192
- date code, 174
- description, 23, 191
- diffusion, *see* diffusion
- electrochemical, 22, 26, 29, 30
- electrolyte, *see* Electrolyte
- failure, 26, 172, 174
- final voltage, 31
- full charge, 31
- gassing, 26, 36, 45, 194
- gassing voltage, 26, 36, 38
- grid, 194
- heat, *see* Heat
- how it works, 22-26
- hydrogen, 25, 27, 29
- hydrometer, 169, 194
- inefficiency, 29
- installation, 56
- internal resistance, *see* Internal resistance
- isolator, *see* Isolator
- life, 33, 167
- maintenance, 53
- open circuit voltage, 34, 148, 171, 194
- overcharging, 26, 174
- oxygen, 24, 27, 29
- parallel, batteries in, 50
- parallel batteries to start engine, 114
- paralleling problems, 51
- plates, 22, 43, 123
- power, 34, 142
- problem as power source, 34, 150
- safety, 53
- selector switch, *see* Selector switch
- self-discharge, *see* Self-discharge
- separators, 22
- series, batteries in, 50
- specific gravity, *see* Specific gravity
- specifications (table 4-1), 49
- state of charge, *see* State of charge
- storage, 55
- sulfate, *see* Sulfate
- sulfation, 27, 40
- sulfuric acid, 22, 24, 25
- surface charge, 32, 77
- temperature, *see* Temperature
- troubleshooting, 167-175
- undercharged, 173
- voltage, *see* Voltage

Battery capacity
- 20 hour rate, 48
- adding, 49, 95
- amp-hour rate, 48
- on boat, 111, 117
- CCA, 27, 47, 192
- CCP, 47
- charging effect, 38
- current absorbed, 36
- definition, 192
- determine, 41
- determine (Chapter 3), 21
- discharging effect, 33
- factors to determine, 43
- guidelines, 94
- for inverter, 11, 15
- MCA, 47
- peak capacity, 47
- ratings, 47
- requirements, 40, 52
- reserve capacity, 47, 195
- on RV, 100, 105
- small batteries, 45
- specifications, 48
- temperature effect, 27
- undercharging effect, 26

Battery chargers, *see* Charging devices

Battery charging, (Chapter 5), *see also* Charging devices
- amperage, high, 39
- on boat, 112, 118
- current acceptance, 36-39
- equalization, *see* Equalization
- gassing, 25, 27, 45, 194
- inefficiency, 29
- loads operating, 96
- overcharging, 26, 174
- practical charging rate, 39
- process, 25, 57
- on RV, 100, 109
- state of charge graph (Fig 5-2), 61
- time required, 39
- undercharging, 26, 173
- voltage, *see* Voltage

Battery discharging
- 50%, to less than, 33
- amperage, high, 34, 150
- definition, 19
- gel batteries, 46
- inefficiency, 29

process, 23, 30
rapid, 32
rate determines capacity, 48
temperature effect, 27
voltage, *see* Voltage

Battery types, 43-46
4D and 8D, 46, 111
chassis battery, 6
coach battery, 6
deep-cycle, 34, 43, 45, 111, 193
gel, 46
house battery, 6
maintenance free, 44
sealed cells, 44
starting battery, 6, 23, 33, 43

Bulk stage, 63, 192

Buss bar
definition, 138, 192
locating, 141

C

Capacity, battery, *see* Battery capacity

Cell, 23, 192

Charging circuits
troubleshooting, 165

Charging Cycle
boat, with generator, 118
boat, with multi-stage alternator, 112
on small RV, 100
RV with generator, 109

Charging devices, (Chapter 5)
absorption stage, 64
alternator, *see* Alternator
battery charger, 67, 101
boost chargers, 67
bulk stage, 63
constant current/constant voltage, 63
constant voltage, 68
converters, 67
equalization stage, *see* Equalization
ferro-resonant, 68
float stage, 64
multi-stage chargers, 62, 69, 105
output, determine required, 73, 95
portable generator, 68, 102
RV, on, 105
solar panels, 71
specifications, 70
taper, 36, 37, 60, 63, 67
trickle chargers, 67
troubleshooting, 165, 185
voltage regulator, *see* Voltage regulator
wind generator, 72

Circuit, *see* Electrical circuits

Circuit breaker, 134, 138, 192

Closed circuit, 6

Connectors, 84

Continuity, 133

Converters, *see* Charging devices

Current
charging effect, 36
definition, 6, 122, 193
discharging effect, 34
measurement, 142
resistance determines, 141
surge, 12

D

Date code, battery, 174

DC usage, estimate, 9

Diffusion
cell, 30
charging effect, 36, 39
definition, 193
discharging effect, 33
temperature effect, 27

Diode
description, 88, 193
isolator, 88
problem with, 88
solar panel, 71
troubleshooting, 180
wind generator, 73

Direct current, *see also* Current
definition, 193

Distribution panel
description, 138
identify, 140

E

Electrical
components, 134
continuity, 133
controls, 134
definitions, *see* the Glossary, 5
device, lists, 9, 13
keys to self-sufficiency, 3, 93

Electrical *(continued)*
- load, *see* Load, electrical
- panel, 81
- power, 126, 142
- power source, 134, 139
- power source, problem with, 34, 150
- requirements, determine (Chapter 2), 7
- resistance, 124
- wiring or conductive path, 134

Electrical circuits
- break in a circuit, finding, 156
- as a circle, 135
- components, 134
- definition, 192
- identifying components, 138
- leakage, 172, 194
- lighting circuits, failed, 164
- loads fail, all, 163
- output of charging devices, 166
- parallel, 136
- polarity, 128, 132
- practice with, 141
- failures, reasons for, 156
- resistance, 141
- schematics, 155
- series, 135
- short circuit, finding, 160
- troubleshooting, *see* Troubleshooting
- understanding (Chapter 8), 121
- voltage drop, find, 162

Electrical system
- boat, with generator, 115
- boat, without generator, 110
- designing (Chapter 7), 93
- keys to self-sufficiency, 93
- recommendations, 99
- RV, small, 99
- RV, with generator, 103

Electricity
- definition, 122, 193
- short course, 121

Electrolyte
- battery failure, 26
- description, 22, 193
- diffusion, 30
- troubleshooting, 168

Electronics
- equalization, during, 65
- measure resistance, 131
- troubleshooting failed devices, 164

Electrons, 23, 25, 122, 193

Equalization
- charging, 109
- description, 65, 193
- sulfation, to remove, 27, 40

F

Float stage, 64, 193

Fluorescent lights, 131

Fuse
- description, 134, 193
- distribution panels, 138
- inverter, 106
- locating, 140
- resistance, measure, 131

G

Gassing, 25, 36, 38, 45, 194

Gassing voltage, 26, 36, 38

Generator, 68, 102, 103, 115

Ground
- definition, 136, 194
- identifying, 139

H

Heat
- battery inefficiency, 29
- battery overheating, 28, 39
- due to resistance, 146

Hertz, 194

Hydrometer, 166, 194

I

Internal resistance
- deep discharging effect, 34, 150
- factors to determine, 29
- sulfate effect, 33, 150
- temperature effect, 27

Inverter, 11-16
- on boat, 117
- current, amount of, 84
- description, 11
- inefficiency, 14, 19
- limitation, 15
- multi-stage charger, 70, 105, 117
- requirements, 12

on RV, 105
sizing of, 13
Ions, 22
Isolator, 88
charging, while, 103
description, 88
locating, 139
problem with, 88
troubleshooting, 180

K

Keys to self-sufficiency, 3, 93

L

Leakage, electrical, 175
Load, electrical
definition, 6, 194
description, 134
locating, 140
power, determine, 142
resistance, 124
Load tester, 172, 194

M

Monitors, (Chapter 6)
ammeters, 78
amp-hour meter, 81
boat, on, 112, 118
electrical panel, 81
multimeter, 126
RV, on, 106
voltmeters, 76
Multi-stage chargers, *see* Charging devices
Multimeter, *see also* Voltmeter, Ammeter, Ohmmeter
analog, 126
continuity check, 133
digital, 126
practice with, 127

O

Ohm, 124, 194
Ohm's law
definition, 125, 194
determines current, 141
determines voltage, 144
Ohmmeter, *see also* Multimeter
resistance measurement, 130-133
Open circuit
definition, 6, 191
voltage readings, 171
Operating your system (Chapter 7), 93

P

Parallel circuit
definition, 136, 194
identifying, 140
Paralleled batteries, 50
problem with, 51
to start engine, 114
Polarity, 128, 132
Portable generators, 68
application, 102
Power
definition, 126
in a load, 34, 142

R

Rate of charge, 6
Rate of discharge, 6
Recommendations, *see* Chapter 7
boat, with generator, 115
boat, without generator, 110
RV, small, 99
RV with generator, 103
Regulator, *see* voltage regulator
Resistance
battery internal, *see* internal resistance
current, determines, 141
definition, 124, 195
measurement, 130-136
temperature effect, 27
unwanted, 144
voltage drop, 145
wiring, 82, 146
RVs and boats are similar, 3

S

Schematics, 155
boat with inverter/multi-stage charger and regulator, fig 7-6, 118
boat with multi-stage regulator, fig 7-5, 114

Schematics *(continued)*
RV with converter, fig 7-2, 101
RV with inverter/multi-stage charger, fig 7-4, 108
RV with multi-stage charger, fig 7-3, 107
Selector switch, 85-87
description, 85, 192
locating, 139
measure resistance, 132
operation on boat, 114
operation on RV, 109
problems, 87
Self-discharge
deep-cycle batteries, 45
description, 30
solar panels preventing, 72
temperature effect, 28
Separators, 22, 195
Series circuits, 135, 195
Short circuits
definition, 195
finding a short, 160
Shunt, 80, 195
Solar panels, 71
application, 96
on boat, 112, 118
description, 71
on RV, 106
troubleshooting, 189
Solenoid, 90
description, 90
locating, 139
troubleshooting, 183
Specific gravity
charging, during, 26
definition, 23, 195
diffusion, during, 30
discharging, during, 24
readings, 170
sealed batteries, 44
State of charge
definition, 6, 195
graph (Fig 5-2), 61
measuring electrolyte, 23
measuring voltage, 32, 76
rate of charge effect, 38
readings, 170, 171
time to restore, 62
Sulfate, 23, 25, 26, 33, 34
Sulfation, 27, 40
Sulfuric acid, 22, 24, 25
Surface charge, 32, 77
Switch, *see* Selector switch

T

Taper chargers
description, 60, 67
Temperature
batteries, effecting, 27, 34, 170
diffusion effect, 27
self-discharge effect, 30
specific gravity, and, 23
Test Light
AC type, 129
DC type, 128
description, 127
Trickle charging, 27, 67
Troubleshooting, (Chapter 9)
alternator, 177-180
battery, 167-177
battery, overcharged, 174
battery, undercharged, 173
battery chargers, 185-189
battery fails to keep a charge, 172
battery failure, determine cause of, 174
break in a circuit, finding a break, 148, 156
charging circuits, 165
electrical leakage, 175
electronic devices, 164
failures, reasons for, 156
isolators and diodes, 180
lighting circuits, 164
loads fail, all, 163
RV converter/battery charger, 187
short circuit, finding a, 160
solar panels, 189
solenoid, 183
voltage drop, 162
wind generators, 190

U

Unwanted Resistance, 144

V

Voltage
charging, 36, 77
charging effect, 37
circuit voltage, measure, 146
closed circuit, 147,151
constant current, 62, 63
constant voltage, 60
definition, 6, 122, 195
difference in potential, 122
discharge, during, 30
discharging effect, 33, 34
drop, 35, 144, 148, 162, 195
final voltage, 31
gassing voltage, 26, 36
load under, 77
measurement, 127, 128
open circuit, 6, 34, 76, 148, 171
range, 31
readings, 76
state of charge graph (Fig 5-2), 61
temperature effect, 27
Voltage regulator
absorption stage, 64
automobile, 60
bulk stage, 63
constant current, 62, 63
constant voltage, 60
description, 59
equalization, 65
float stage, 64
multi-stage, 66
problem with automobile regulators, 61
rewiring, 89, 181
wind generator, 73
Voltmeter, *see also* Multimeter, Monitors
description, 76
electrical panel, 81
installation, 78
voltage measurement, 127, 128
voltage readings, 76

W

Watts
definition, 196
explanation, 126
formula, 8
Wind generator, 72
application, 96
boat, on, 118
description, 72
troubleshooting, 190
Wiring
AWG, 83
conductive path, 134
description, 82
locating, 140
resistance, 82, 124, 146
Worksheets
battery capacity, determine, 40
charging device, determine output of, 73
electrical requirement, determine, 17

Harold Barre has been a naval officer, a manager at a major Silicon Valley electronic company and in boating for 20 years. For 14 years, he has lived, using 12 volts, onboard a sailboat and sailed for 4 of those years from California to Hawaii, through the Panama Canal, to Baltimore, and Martinique. In an RV, he has traveled throughout the United States and Canada to Alaska and Mexico, so he understands what a boater or an RVer needs to know to manage successfully a 12 volt electrical system.